Beyond
NIMBY

Beyond NIMBY

Hazardous Waste Siting in Canada and
the United States

Barry G. Rabe

THE BROOKINGS INSTITUTION
Washington, D.C.

About Brookings

The Brookings Institution is a private nonprofit organization devoted to research, education, and publication on important issues of domestic and foreign policy. Its principal purpose is to bring knowledge to bear on current and emerging policy problems. The Institution was founded on December 8, 1927, to merge the activities of the Institute for Government Research, founded in 1916, the Institute of Economics, founded in 1922, and the Robert Brookings Graduate School of Economics, founded in 1924.

The Institution maintains a position of neutrality on issues of public policy. Interpretations or conclusions in Brookings publications should be understood to be solely those of the authors.

Copyright ©1994
THE BROOKINGS INSTITUTION
1775 Massachusetts Avenue, N.W., Washington, D.C. 20036

Library of Congress Cataloging-in-Publication Data:
Rabe, Barry George, 1957-
   Beyond Nimby: hazardous waste siting in Canada and the United States / Barry G. Rabe.
      p. cm.
   Includes bibliographical references and index.
   ISBN 0-8157-7308-0 — ISBN 0-8157-7307-2 (pbk.)
   1. Hazardous waste sites — Location — Political aspects — Canada.
2. Hazardous waste sites — Location — Political aspects — United States. I. Title
TD1050.P64R33 1994
363.72'87'0971 — dc20   94–19039
                                                                        CIP

9 8 7 6 5 4 3 2 1

The paper used in this publication meets the minimum requirements of the American National Standard for Information Sciences — Permanence of paper for Printed Library Materials, ANSI Z39.48-1984

Earlier versions of portions of this book have appeared in *Governance: An International Journal of Policy and Administration* (April 1991), *Journal of Health Politics, Policy and Law* (Spring 1992), *Journal of Resource Management and Technology* (March 1994), and *Environmental Law*, no. 1 (1994).

Set in Elektra

Composition by Graphic Composition, Inc.
Athens, Georgia

Printed by R. R. Donnelley and Sons Co.
Harrisonburg, Virginia

For my parents,
George and Marian Rabe

Contents

Preface   xi

1. The Politics of Hazardous Waste Facility Siting   1

Toward a Voluntary Approach to Facility Siting   4
The Pursuit of Cooperation   6
A Problem Shared   10
The Costs to Society of Siting Gridlock   13
Signs of Convergence   22
Factors Fostering Convergence   26

2. When Siting Does Not Work   28

Market Approaches: The Limits of Compensation   33
Regulatory Approaches: The Limits of Preemption   44
The Common Nimby Outcome   56

3. The Alberta Case   58

Beyond the Nimby Syndrome in Alberta   61
Alberta and Its Waste Management Problem   62
System Performance   86

4.   Prospects for Replication   90

Manitoba: Alberta Redux   91
North Carolina: The Greensboro Exception amid Nimby
     Chaos   106
Minnesota: Voluntarism, Nimby, and a New
     Commitment to Prevention   110
California: Promise for Burden Sharing Undermined by
     Bureaucratic Objection   118
Quebec: Siting Success, Implementation Debacle   121

5.   Low-Level Radioactive Waste   128

Early Stages of LLRW Management in Canada and the
     United States   129
The Evolution of Canadian LLRW Management   131
The Evolution of American LLRW Management   140
An Attempt at Voluntarism   143

6.   Toward a More Mature System   148

Canada: The Virtues and Pitfalls of Pure
     Decentralization   150
The United States: The Mixed Record of Conjoint
     Federalism   152
Policy Options for the 1990s and Beyond   153
Environmental Regulatory Integration   165
Looking Beyond Waste   167

Notes   171

Index   195

Tables

1-1. U.S.-Canada Hazardous Waste Trade, 1987–90   20
5-1. Low-Level Radioactive Waste (LLRW) Policy
     Amendments Act Deadlines   141

Figures

1-1. Likelihood of Facility Siting Agreements under
     Alternative Strategies   5
2-1. Approaches to Hazardous Waste Facility Siting Policy in
     Canada and the United States   30

Preface

WORK ON THIS BOOK began in 1979, when I was enrolled in an undergraduate political science seminar at Carthage College in Kenosha, Wisconsin. The professor, William Gunderson, introduced his five students to diverse theoretical lenses through which to examine public policy. For our final paper, we were asked to test what we had learned against a specific case of policy conflict. I chose an episode that was transpiring in my hometown, a middle-class suburb twenty miles west of downtown Chicago. The case involved a proposed zoning change, which sought to convert a nursing home into a nonprofit center for alcohol treatment. The building in question had initially been a single-family residence and had suffered from some disrepair. The proposed facility, called a detox center, was designed to serve individuals from my hometown and surrounding community. Plans called for some renovation of the building and strict limits on the number of patients allowed on the premises at any one time.

The case intrigued me for the simple reason that the public announcement of this facility conversion triggered a level of political outrage unlike anything I had ever experienced in the town in which I grew up. Petitions were filled with signatures, hearings of the normally obscure zoning commission were packed, hostile signs sprang up along dozens of lawns, and a number of elected officials (and aspirants) took vocal positions on the matter. All of this activity was concentrated into a vigorous effort to force the detox facility to go somewhere—anywhere—else. In the numerous interviews that I conducted, I heard no one object to the idea of such a facility or contend that no need existed for such services for some residents of our community. But none of the opponents wanted any facility anywhere near his or her home. The most commonly cited reasons for opposition were the economic fears of declining property values and the safety risks of having people with drinking problems about in the neighborhood.

Facing such opposition, as well as a series of threatening letters and phone calls, the managers of the proposed facility threw in the towel. The detox center would attempt to go elsewhere, although it encountered comparable problems in other communities. In writing my research paper, I tried to explain what had happened and reflect on how such a needed but controversial facility might best be placed somewhere. The paper was incomplete at best, but I remained fascinated by the case and hoped I could revisit such an issue at some future point.

Then came the not-in-my-backyard, or Nimby, syndrome. About a decade after my initial foray into Nimby politics, I selected a larger target: the siting of hazardous waste facilities in the United States and Canada. Episodes such as kepone releases along the James River, and the Mississauga, Ontario, train derailment and its potential release of toxic materials; and such discoveries as the Valley of the Drums and the Love Canal put hazardous waste on the agendas of national and subnational governments in both countries. Existing technologies for disposal were highly suspect, and massive quantities of abandoned wastes were discovered that required treatment. Given the ubiquitous nature of these wastes, they were unlikely to cease being generated, barring dramatic deindustrialization of North America. The most ambitious plans for reduction in the volumes of new waste being generated would not eliminate the larger problem, even if they were fully implemented. New, more sophisticated facilities would need to be created. But in beginning to examine this issue, I learned that state and provincial governments had been about as successful in opening new waste management facilities as detox center proponents had been in my hometown.

Cases of siting gridlock were easy to find and followed similar patterns in respective states and provinces. In the absence of prior consultation, communities were confronted by either governmental agencies or private waste management firms with the news that they had been selected to host a major waste disposal facility. These communities were outraged and repeatedly took such swift and decisive opposing action that agencies and firms eventually withdrew their proposals. This inability to open new facilities had some salutary effect, including intensified pressure for waste reduction and rejection of a number of poorly conceived facilities. But, in many respects, these cases contributed to growing distrust over every aspect of hazardous waste management and delayed the development of a viable, long-term management approach worthy of public support.

As I surveyed this messy landscape, I chose to focus on alternative approaches to siting and waste management that deviated from the familiar patterns. They were not easy to find. The rapidly growing literature on the politics of waste facility siting and management offers tale after tale of public outrage and siting gridlock. Similarly, the growing literature on environmental policy tends to pinpoint what has gone wrong in the past but gives little clear sense of where to head next.

After sifting through nearly two dozen cases, I was struck to find some deviations from the Nimby norm. When conditions are met that differ from standard approaches to siting and waste management, cooperation becomes possible. Those conditions and the cases in which they were applied provide the central focus of this book. In Alberta, Canada, for example, a comprehensive system of hazardous waste management is now in place and is publicly supported. This system actively promotes maximal waste reduction and also provides a network involving community drop-off stations, regional transfer facilities, a sophisticated network for waste transportation, and a comprehensive central waste management facility. This system was created through an open political process, with siting proceeding only after extensive public dialogue and formal community acceptance of new facilities. In turn, the province has taken far-reaching steps to protect host communities from exploitation and assure safe, long-term facility management.

Alberta, it turns out, is not a fluke. Other provinces and states have devised their own variants on this approach, with some impressive results to date. These cases do not add up to some definitive theoretical assertion that such an approach guarantees siting agreements and political concord. But given the enormous political energies and economic resources devoted to the struggle over siting in American and Canadian settings, they offer an alternative to the prevailing, combative order. I do not know if such an approach might be applicable to other areas where Nimby conflict is rampant, such as nursing homes, prisons, hospices, public housing, and drug and alcohol treatment centers. But having seen so many cases so similar to the one I encountered in 1979, with such disturbing results, I think it warrants serious consideration.

The Plan of the Book

I examine the types of siting policies for hazardous wastes and their implementation processes in a diverse range of states and provinces. Chapter 1 provides a review of the ongoing problem of waste management in North
America. It introduces the main regulatory and intergovernmental features
of facility siting in Canada and the United States and discusses the policy
ramifications of the Nimby syndrome. Unlike many works in comparative policy analysis that accentuate differences in styles of national policymaking and
implementation, this book emphasizes that strong similarities exist between
Canadian provinces and American states in hazardous waste facility siting and
management, making their joint consideration particularly fruitful. Moreover,
extensive waste trading occurs across national boundaries, underscoring the
need for effective binational policy. The special focus of chapter 2 is the familiar gridlock that emerges in most siting cases. I analyze a series of American
and Canadian cases that employed nearly identical siting strategies and resulted in the same Nimby conclusions. Instead of dwelling on these cases of
conflict, which fall into fairly predictable and easily explainable patterns, I
devote much attention in chapters 3 through 5 to an analysis of the conditions
that must be met for siting agreements to be reached. I also discuss the related
area of low-level radioactive waste facility siting and the process whereby one
Canadian province and one American state have made considerable progress
toward siting. The concluding chapter draws broad lessons from the experiences of successful siting, explores policy options for Canada and the United
States, and considers the prospects for applying the findings of this study to
other areas of domestic policy in which Nimby conflict is common.

The emphasis on the political aspects of facility siting is intended to supplement the dominant disciplinary forces in this policy area. On the one hand,
much hazardous waste policy has been shaped by economics, which has fostered the perception held by many governmental and corporate officials in
both nations that market mechanisms could resolve any waste facility siting or
waste management problems in a timely, efficient manner. Unfortunately, the
evidence indicates that economic strategies, including those that emphasize
generous compensation packages as the ultimate balm of siting tensions, almost invariably fail. On the other hand, a good deal of hazardous waste policy
has been influenced by engineering and operations research, which has attempted to devise precise, technical criteria for selecting preferred sites. Public and private officials in both nations have assumed that such criteria would
lead them to technically ideal areas for siting and that any problems could be

resolved through the side payments proposed by economists or by regulatory strong-arm tactics. But time and again, this approach has resulted in political turmoil and siting gridlock.

This book attempts to step beyond these limitations with a fairly simple strategy: explaining the familiar pattern of Nimby conflict and then developing a framework to explain the common conditions under which siting agreements might be reachable, based on those exceptional cases in which cooperation did emerge. It heeds the call of political scientist Elinor Ostrom, who in *Governing the Commons: The Evolution of Institutions for Collective Actions* contended, "Given the presumption of failure that categorizes so much of the policy literature, it is important to present examples of success." In so doing, this book continues a branch of social science inquiry endorsed by sociologist James S. Coleman, the use of comparative case studies to better understand social and political conflict. In the 1950s Coleman analyzed a series of highly conflictual cases in the United States, including school desegregation, anti-Communist sentiment, and water fluoridation. None of these is directly analogous to the current Nimby phenomenon in Canada and the United States, although some of the strategies used to defuse controversy are strikingly similar to those employed successfully in select siting cases of the 1980s and 1990s. Coleman, in his 1957 book *Community Conflict*, noted the common themes in the cases he studied but cautioned that "these controversies are not determinate things which grind their way unswervingly to the end. They may take a number of courses; the outcome may range from amicable resolution to bitter name-calling. Modifying factors which differ among communities may intensify controversy or reduce its tensions." Much of this book will explore factors that may contribute to greater cooperation, placing particular emphasis, as did Coleman, on the issue of finding constructive mechanisms of public participation in the policymaking process.

Acknowledgments

In undertaking this odyssey into the politics of facility siting and alternatives, I have incurred countless debts. More than 150 individuals consented to be interviewed for this project, some repeatedly, and their insights were essential to the development of the case studies. Financial support from the Canadian-American Committee of the National Planning Association and C. D. Howe Institute, the Canadian Studies Faculty Research Grant Program, and the Joyce Foundation was essential in pursuing field work and other

aspects of the project. In particular, this support allowed me to hire, at vary-
ing stages of the project, a series of superb research assistants: Beth Caretti,
Richard Compton, Debbie Cozans, Margaret Daniel, Laura Flinchbaugh,
Hilary Frazier, Beth Lowe, Jessica Miller, Robin Norton, Marion Perrin,
Pam Protzel, and Beth Weinberger.

I have also benefited greatly from the ideas and suggestions of numerous
colleagues during various states of the manuscript's development. Nancy D.
Davidson encouraged me to write this book and advised me on how to im-
prove it once it was drafted. Charles F. Doran, Doris Dunsmore, William T.
Gormley, Janet Ketcham, Jonathan Lemco, James A. Morone, Leslie A. Pal,
and Eric M. Uslaner read the entire manuscript and offered extremely useful
comments. In addition, Odin Anderson, David Caffry, Colin Campbell,
Geoffrey Castle, Toby Citrin, Thomas A. D'Aunno, John M. Gillroy, Frances
Irwin, Robert A. Katzmann, Donald F. Kettl, Anne M. Khademian, Michael
E. Kraft, Karl Kronebusch, Douglas Macmillan, Daniel A. Mazmanian, Paul
Muldoon, Philip A. Mundo, Donald Munton, Paul E. Peterson, Paul J. Quirk,
John H. Romani, John Samples, Mark Schneider, Mildred A. Schwartz, Syl-
via N. Tesh, John T. Tierney, Konrad von Moltke, Kathy Wagner, Kenneth E.
Warner, and Janet B. Zimmerman read portions of the manuscript or discussed key themes, some at very early stages, and assisted greatly in its refinement. William C. Gunderson and Peter T. Harbage not only contributed
to my general understanding of Nimby and facility siting but also served as
coauthors of chapter 5. Colleen McGuiness provided valuable advice on style
and content in the course of editing the manuscript for publication. Becky
Pace played a central role, serving as an exemplary secretary throughout the
course of the project and as verifier in the final stages. Many thanks also to
numerous students at the University of Michigan and the University of
Wisconsin-Madison who sharpened my thinking about these issues.

My efforts were made far easier by the comfortable, supportive venues I
have enjoyed in writing. Most of the early drafts were written in the friendly
confines of my University of Michigan escape hatch, the Matthaei Botanical
Gardens, and I am most grateful to Erich E. Steiner, Patricia S. Hopkinson,
David C. Michener, Susan H. Boss, and all the members of the Gardens staff
for their many kindnesses throughout. Various revisions were completed in
other locales. During the fall of 1993 I was a visiting fellow at the Institute of
International Environmental Governance at Dartmouth College, where Jean
Hennessey, Nicki Maynard, and Oran Young provided a comfortable, colle-
gial setting in which to write. During the winter of 1994 I completed final
revisions while a visiting associate professor at the Robert M. La Follette Insti-

tute of Public Affairs at the University of Wisconsin-Madison. I remain most grateful to Peter Eisinger, director of the institute, and other Wisconsin colleagues for their support, stimulation, and good humor during this stage.

I would also like to acknowledge the people most important in my life. Two of them, sons Matthew and Andrew, were not yet born when this project began. They are a wonderful source of energy and love, as well as a constant reminder that the academic proclivities for workaholism must be resisted at every step. I would also like to thank Matthew for his steady support of artwork related to "Dad's book on pollution." My wife, Dana Runestad, has contributed to this book and my life in ways that defy description, and she has remained my closest friend throughout this project. Finally, I want to publicly thank my parents, Marian and George Rabe, to whom this book is lovingly dedicated. They have been a steadfast source of support in all my endeavors, including this one.

Beyond
NIMBY

1

The Politics of Hazardous Waste Facility Siting

NO NATION spends as much money per capita, generates as much political anguish, or accomplishes less in implementing its hazardous waste policies than does the United States. Few nations fall as close to the United States in these respects as does Canada. The politics of hazardous waste facility siting, and hazardous waste management more generally, thus constitute a fundamental political challenge for these neighboring North American nations.

In both settings, hazardous waste poses a classic political problem of regulatory cost distribution. Waste generation occurs throughout North American states, provinces, and territories. Precise definitions and measures vary, but a generally accepted statistic is that a little more than one ton of hazardous waste is generated each year for each North American citizen. Although much of this is concentrated within industry, the average household generates approximately fifteen pounds of hazardous waste each year. Estimates may increase in future years, given the expansive definition of what constitutes hazardous waste, as opposed to solid or other types of waste. Any long-term strategy for regulating waste thus involves more than targeting a handful of large waste generators for remedial actions.

The benefits of the activities that produce massive quantities of wastes are numerous and broadly distributed. These range from an economic base that remains enviable by international standards to unsurpassed consumer choice and convenience. Politicians interfere at their peril, given their general predilection for actions that enable them to distribute popular goods ("claim credit") or that at least are unlikely to generate controversy ("avoid blame").[1] But waste, especially tens of millions of tons of it, must go somewhere, whether to be burned, dumped, or treated. Therein lies the political rub of hazardous waste facility siting. By placing the responsibility for disposal and treatment on those communities that host large facilities, most of the economic, political, environmental, and public health burden imposed by hazardous waste will be concentrated on a small number of communities. Of all the challenges that contemporary policymakers must confront, few can be more difficult than those that narrowly concentrate costs.[2]

Recent experience in the United States and Canada has shown that most communities and their elected officials are less than enamored of the idea of assuming primary responsibility for hazardous waste. They have ample motivation to take political action in trying to oppose facility siting on — or anywhere near — their home turf. Almost without fail, decisive action is taken swiftly once a siting proposal is revealed. Tactics may vary, ranging from litigation to picket lines, but they are almost always effective in thwarting proposals. As a result, virtually no new hazardous waste treatment or disposal facilities have been opened in either nation since the mid-1980s.[3] The prospects for adding treatment, disposal, or recycling capacity appear grim in the years ahead under prevailing policy. Furthermore, many extant facilities that were opened in earlier periods before hazardous waste became so controversial are increasingly unpopular, particularly those that handle large quantities of waste from other states or provinces.

A new acronym has been ushered into the North American political lexicon: Nimby. The not-in-my-backyard syndrome reflects the growing penchant to block an unpopular facility proposal and thereby leave elusive a long-term strategy for hazardous waste management. An enormous quantity of ink has been spilled by academics and the popular press in attempting to define and analyze this phenomenon in recent years. In the eyes of some, Nimby constitutes an unabashed pursuit of narrow self-interest, whereby "chemophobes" and environmental fanatics try to bring industrialized economies to their knees. In the eyes of others, Nimby is a triumph of Western democracy, as virtuous citizens band together in search of political and environmental justice and usher in an era of "ecodemocracy."

This book rejects both interpretations, instead seeing Nimby as a realistic local response to an immediate problem: national and subnational policies that were poorly designed and ham-handedly implemented. Time and again, officials representing states, provinces, and waste management corporations have eschewed efforts to foster serious public deliberation over the problems posed by hazardous waste and exploration of equitable distribution of burden for treatment and disposal. Instead, they have tended to rely on economic and technical criteria to designate preferred sites. Upon completing such calculations, they assume that siting will proceed either through strong-armed regulatory tactics (preemption) or bargaining that emphasizes economic enticements for cooperation (compensation).

From Florida to British Columbia, both approaches have resulted in repeated rejection of siting proposals for more than a decade. They have also contributed to a general atmosphere of cynicism toward public and private institutions that play some direct role in hazardous waste management in the United States and Canada. So much has gone wrong in designing supportable and implementable public policy in this area that envisioning the next step may be difficult.

This book constitutes an effort to begin to chart that future course. Some attention will be devoted to policy failure and its near-term ramifications, while primary consideration will be given to those exceptional cases in which the anticipated Nimby outcome was replaced with a cooperative siting agreement. In at least a few instances, policy modification has resulted in the creation of serious public deliberation and ultimate political support for facility construction and operation. The most notable of these are comprehensive waste treatment and disposal facilities approved in the Canadian provinces of Alberta and Manitoba. Many components essential to cooperation are also evident in a few of the most promising hazardous waste facility siting experiments in various American states and other Canadian provinces, as well as examples from the equally contentious area of siting low-level radioactive waste disposal facilities.

The analysis of cases that defy the conventional Nimby pattern will not lead to any easy recipe or set of blueprints that public or private sector officials can follow. It begins with a call for more talk—with the intent of providing citizens and communities with essential information and direct opportunity to shape policy—instead of a push for rapid construction of multiple facilities. But it will hold out considerable hope that siting gridlock and the attendant economic, political, and environmental problems need not be the only possible outcome.

Toward a Voluntary Approach to Facility Siting

The cases examined in this book reveal that subnational governments in both Canada and the United States must not only respect and protect the political choices of their citizens but also must encourage and facilitate their participation in every dimension of the siting process and ultimate facility operation, thereby providing constant reassurance against the threat of exploitation. Individuals must be able to trust the government as a representative of the common interest in determining what is a collective good if siting is to occur and subsequent facility operation is to retain public support.[4] This constitutes a fundamental challenge to the more traditional approach to siting, a top-down process in which either governmental or corporate officials engage in little if any prior consultation with citizens and communities before declaring siting targets.

Before siting proposals are offered or the actual siting takes place, the government must begin a dialogue with its citizens about the general parameters of the hazardous waste problem that may make the siting necessary, the dynamics of the problem, and the range of possible solutions.[5] Consideration of potential hosts is appropriate only after full participation and explanation of the necessity and risks of the siting. Individuals presumably want to be respected in their choice to accept or reject a risk to themselves and their community. Full disclosure of relevant information is required, as well as the assurance that siting will take place only upon some formal demonstration of community trust and acceptance, such as a local referendum. This type of process is designated as "voluntary" in figure 1-1, contrary to the more commonplace "coercive" approach to siting that occurs in the absence of prior consultation and often involves efforts to entice or compel communities to cooperate.

The promotion of greater dialogue in siting, however, is insufficient to create an atmosphere of trust. It needs to be supplemented with further assurances that any community contemplating the hosting of a facility will not become the sole dumping ground for long-term waste management for a broad geographic region. Throughout any siting process, it must also be made evident that the construction and operation of a facility is only part of a multifaceted provincial, state, or regional strategy for waste management and that any host community will not be alone in working toward hazardous waste policy that will protect the environment and safeguard public health.

Each new site must be established in full knowledge of its importance and its risks but with confidence that free riders will be controlled. Any new facility

Figure 1-1. *Likelihood of Facility Siting Agreements under Alternative Strategies*

	Waste management burden concentrated on site host	Waste management burden shared widely with other communities, states, or provinces
Voluntary siting process	Agreement very unlikely 1	Agreement possible 2
Coercive siting process	Agreement impossible 3	Agreement very unlikely 4

must be linked with far-reaching waste recycling and waste reduction initiatives as well as regional facilities for sorting wastes and preparing them for transfer to the disposal facility, thus distributing broadly the responsibility for waste management instead of thrusting all of it upon a single community. Furthermore, prospective host communities must be guaranteed that their unilateral cooperative action will not be exploited; that is, they will not be made magnets for wastes generated in states and provinces unwilling to share in the burden of waste management and develop their own disposal capacity. This does not necessitate lodging National Guard troops or Canadian Mounties on subnational or national boundaries to block waste shipments but does indicate the need for formal burden-sharing arrangements among multiple units of government. In the traditional pattern of Canadian and American siting episodes, the burden-sharing features of the siting process are commonly ignored.

In all of the successful siting agreements discussed in this book, a combination of a voluntary siting process and formal commitment to burden sharing was evident. As reflected in cell 2 of figure 1-1, siting agreements become a real possibility under these circumstances. By contrast, the prospects for a siting agreement diminish markedly if either of these two features is not included (cells 1 and 4) and disappear if both are absent (cell 3).

If citizens are confident that their provincial or state government — and any corporate waste management partners — represents the public interest in providing necessary collective goods, and they are allowed full participation in the myriad decisions that must be made before a facility site is chosen, then they will not feel backed into a corner. Thus they will be free to function without the pressures of fear, uncertainty, and isolation so common in most siting cases in Canada and the United States. They can seriously and confidently weigh various economic and social compensation packages that may accompany a siting agreement, unlike traditional siting approaches, which assume that compensation alone will entice cooperation. They can consider the problem as a collective one in which they are intimately involved in the decisions and privy to all necessary information. Recent experience in hazardous waste facility siting in Canada and the United States suggests that this may be the only way to begin to move beyond siting gridlock and toward a more mature system of waste management.

The Pursuit of Cooperation

Recent research in political science and public policy indicates that a transformation of siting policy is difficult but not impossible. Much political science on both sides of the forty-ninth parallel remains focused on the foibles of national and subnational governments, in particular the difficulty of assuring cooperation among respective units of government and between the public and private sector. By contrast, an understanding of "when government works" remains highly limited.[6] As political scientist Paul J. Quirk noted, political science has generally failed "to identify the conditions for cooperative resolution of policy conflict."[7] This assessment is especially applicable to the literature that examines environmental politics and policymaking in the two nations, including a good deal of the analysis of hazardous waste policy. Perhaps the greatest contributions in this area to date by political scientists have been through studies of political conflict and policy failure.[8]

Important developments within political science in recent years have begun to explore the conditions necessary for cooperation and for the formation of effective policy and have confirmed that even thorny redistributive problems such as facility siting need not be permanently mired in political conflict. Four key themes emerge from an examination of multiple case studies of siting attempts that have been totally successful, partially successful, or totally unsuccessful. These findings, in combination, point to a rejection of traditional, top-down siting approaches in favor of more democratically based alternatives that concentrate both environmental rights and responsibilities in the hands of individual communities, provinces, and states.

First, overwhelming evidence exists that dominant siting strategies employed in Canada and the United States are magnets for distrust, political adversarialism, and, ultimately, siting gridlock. These strategies provide for a private firm (in the form of a private waste management company) or the public state (in the form of a governmental agency or regulatory body) to select a preferred site and either entice or coerce a surprised community to consent to siting. As political scientist Elinor Ostrom noted, "Both centralization advocates and privatization advocates accept as a central tenet that institutional change must come from outside and be imposed on the individuals affected." [9] Applied to siting, this central tenet is flawed, just as Ostrom finds it wanting in case after case of "common pool" resource protection. In provinces and states with diverse political cultures, communities unexpectedly faced with the prospect of hosting a major disposal facility feel unfairly singled out. They take opposing action of such vigor that the siting proponent almost invariably backs away after a political showdown. Thus, despite the theoretical elegance of pure market or pure regulatory approaches to determining the distribution of siting responsibility, in practice such approaches simply fail to achieve the intended outcome. Over time, they make a difficult situation worse as distrust and adversarialism mount.

Second, significant evidence is available that citizens are capable of engaging in collective deliberation and making reasonable decisions about facility siting and waste management. Both government and waste management firms can play a role in this process, but a fundamental shift occurs with the transference of decisionmaking responsibilities from public and private sector authorities to the general citizenry. The siting process thus begins with democratic discourse that includes consideration of facility need, equitable burden sharing among communities, alternative waste management technologies, and long-term protections. Such discourse takes time and is hardly an elixir sure to eliminate conflict. It is intended to create conditions under which conflict will, in all likelihood, emerge, but participants will be far more likely to consider cooperation and devise effective solutions.[10]

When extended dialogue has preceded site selection, collective deliberation becomes possible and very different siting and waste management outcomes result. This finding is consistent with a growing trend in public opinion research that rejects traditional perceptions of the citizenry as disengaged and preoccupied with its immediate and narrow self-interest. As a central motivation behind political behavior, self-interest is not disregarded, but the general contention is that the citizenry, writ large, can evaluate policy options and welcomes and supports creation of serious participatory opportunities. When such opportunities supplant a priori decisions such as site selection, citizens can and do consider alternative arrangements for waste management. The collective process can have transformative effects on communities as well as government and private experts. In some instances, outcomes are reached that could not have been envisioned at the outset of the deliberative process. In virtually all instances, the collective process results in greater consensus and a more promising strategy for waste management than that achieved under traditional siting approaches.

Third, the shift toward earlier and more extensive public input into waste policy decisions does not occur independently of institutions that will play a key role in implementing any decisions reached. Collective dialogue over hazardous waste does not occur in automatic response to a groundswell of policy expression by the citizenry. Creative institutions, primarily from the public but also at times from the private sector, consistently play a role in establishing a process and forum for deliberation. In each of the cases, from both nations, when traditional Nimby gridlock was replaced with more cooperative outcomes, such institutions emerged to encourage extensive dialogue before the reaching of any essential policy decisions.

This scenario is consistent with a growing body of scholarship that accentuates the role of institutions in the policy process and the constructive contributions that they can make. Institutional analysis is hardly new to disciplines such as political science. Nonetheless, in more recent years scholars have begun to define these institutions more broadly and examine their varying features with greater sensitivity to likely policy outcomes. As organizational theorists James G. March and Johan P. Olsen observed, "By providing a structure of routines, roles, forms, and rules, political institutions organize a potentially disorderly political process. By shaping meaning, political institutions create an interpretive order within which political behavior can be understood and provided continuity." [11] In those siting cases in which institutions emphasize facilitation of political dialogue instead of imposition of final decisions, a considerable capacity for creation of such an interpretive order can be seen. [12]

Institutions such as governmental agencies thus have a role that is very much at odds with much dominant thinking about them. Varying camps of scholars have portrayed these agencies as either lackeys captured by the industries they are supposed to oversee or regulate, overzealous bodies eager to maximize their resources and powers, or entities highly responsive to the manipulation of elected principals, such as legislators or executives. [13] Considerable evidence exists to corroborate these distinct interpretations in the evolution of American and Canadian hazardous waste and environmental regulatory policy.

Nonetheless, alternatives to these dominant understandings of institutional performance are available both in theory and in practice. Institutions and the policy professionals that serve them devised systematic methods to maximize the likelihood of meaningful public dialogue that could then lead to some consensus concerning siting and waste management. The professionals earned widespread public respect and opened up a process in which democratic dialogue could directly contribute to formation and implementation of policy.

Fourth, the general approach to siting endorsed in this book may serve as a broader model to better recast American and Canadian debates over the future course of environmental policy. Most present debates transpire at high decibel levels; opposing camps tend to take extreme positions and give little ground, whether the debate takes place in a public hearing, legislative chamber, or courtroom. In turn, the policymaking process often tends to respond to the loudest voices — or the most dramatizable event — rather than reflect a careful weighing of alternative problems and strategies through extended public discourse. Such patterns are not unique to environmental policy but may be especially common in it, because of the high salience of the involved issues and the conflict engendered in distributing responsibility for remedial or preventive actions.[14]

The alternative approach offers a path that might better balance the environmental rights to which individuals and communities are entitled with their responsibilities in collectively assuring environmental well-being. Much as legal scholar Mary Ann Glendon has noted about other spheres of domestic policy, environmental policy — and hazardous waste facility siting in particular — is laden with "rights talk" but short on discussion of collective responsibility. In siting, one repeatedly hears firm expression of a constituency's rights: government officials feel that their expertise should override community opposition to proposed siting; private waste management firms feel that their right to make a profit and, in some cases, ownership of the property upon which a facility would be established entitles them to operate independently of public sentiments; community groups are quick to assert their rights to reject local siting proposals, even though this may merely deflect waste management burdens elsewhere. As Glendon lamented, "Unfortunately, American political discourse has become vacuous, hard-edged, and inflexible just when it is called upon to encompass economic, social, and environmental problems of unparalleled difficulty and complexity."[15]

Rigidity need not be the only plausible approach to hazardous waste facility siting and environmental policy more generally. Experimentation with so-called alternative dispute resolution for various environmental conflicts in the

1970s and 1980s included site-specific mediation of environmental disputes, negotiation among contending parties before agency rule promulgation, and broad efforts to foster policy dialogue on large issues before formal and official exploration of them. A number of these efforts have been fairly successful, particularly in resolving more narrowly focused conflicts to the satisfaction of all combatants.[16] Larger dialogues, such as the national coal policy project of the late 1970s, have been less effective, both because of their advisory natures and the absence of direct participation by governmental agencies that would play a role in implementing any agreements that were reached.[17]

The siting approach outlined in this book offers a promising variant on these approaches. More than merely attempting to split differences or settle squabbles, it is designed to engage multiple communities within a polity in serious democratic dialogue. More than merely getting differing parties to talk to one another, it offers a way to examine respective environmental rights and responsibilities and consider mechanisms to share responsibility for waste management in an open, equitable manner. Applied more widely, it could contribute to a more mature era of environmental governance. These findings suggest that conflict and policy failure need not be inevitable outcomes, even in as contentious an area as hazardous waste facility siting. They confirm many of the main themes evident in the cases of siting success and point to the beginnings of a more cooperative and effective approach to hazardous waste management in Canada and the United States.

A Problem Shared

Hazardous waste defies precise scientific definition, exact estimation of public health risk through various routes of exposure, or technological agreement on the safest methods for disposal, treatment, or recycling. The word *hazardous* is generally synonymous with the word *toxic* in Canada and the United States. Individual provinces and states tend to define hazardous, as opposed to solid, radioactive, or biomedical wastes, in somewhat differing ways. Over time, however, the definition provided by Congress in the Resource Conservation and Recovery Act (RCRA) of 1976 has become dominant in both nations. The legislation defines hazardous waste "as a solid waste, or combination of solid wastes, which because of its quantity, concentration, or physical, chemical, or infectious characteristics, may cause, or significantly contribute to an increase in mortality or an increase in serious irreversible, or incapacitating reversible, illness; or pose a substantial present or potential hazard to human health or the environment when improperly treated, stored, transported, or disposed of, or otherwise managed." [18] Solid wastes may be deemed hazardous under RCRA if they exhibit one or more of the following characteristics: ignitability, corrosivity, reactivity, or toxicity.[19] At present, hazardous wastes include a wide array of materials, such as acids, metals, sludges, and solvents.

The appropriate classification of various wastes into hazardous or other categorizations remains a major debate in Washington, D. C., Ottawa, and numerous provincial and state capitols. Considerable political pressure has been brought to bear to err on the side of inclusivity, which may add significantly to the list of wastes to be designated hazardous in the years ahead. For example, the U.S. Supreme Court decided in May 1994 that the substantial quantities of ash created by municipal incinerators, which have traditionally been managed as solid waste, must be handled as hazardous waste in the future.[20] Individual states, in turn, are beginning to use federal definitions as the minimum, open to their own expansions. In 1993, for example, Minnesota established timetables whereby an array of commonly used products, including motor oil and oil filters, brake, power-steering and transmission fluids, antifreeze, and fluorescent lamps, may no longer be disposed of as a solid waste.[21] Instead, if not recyclable, they must be taken to hazardous waste disposal facilities. Given the changing definitions, hazardous waste volumes may stabilize or increase even if significant waste reduction and recycling are achieved.

The classification systems for measuring the volumes of these wastes and their toxicity have improved in recent years, particularly in the United States. Most recent analyses indicate that between 275 million and 380 million tons — or at minimum 1 ton per person — are generated each year in the United States. The U.S. Environmental Protection Agency (EPA) in 1989 estimated that hazardous waste volumes created by the largest 20,233 generators were 197.6 million tons, a drop from prior years. However, sizable state-by-state data discrepancies suggest that this estimate is low. In addition, the figure does not address the waste generated by the 250,000 or more so-called small generators or the backlog of wastes awaiting treatment on abandoned waste dump sites. No comparable estimate exists for Canada, in part because of differences in waste generation reporting requirements. Most experts acknowledge a roughly comparable amount of hazardous waste generated per person each year, given the strong similarities in industrialization and consumer practices between the nations. None of these estimates addresses the other aspect of hazardous waste management; how to treat, dispose, or recycle the many millions of tons of wastes generated in prior decades that were dumped indiscriminately — the Superfund sites — and will require some form of cleanup.

Provincial and state governments in Canada and the United States dominate siting policy because of the absence of federal siting legislation in both nations. Subnational authority remains somewhat more dominant in Canada, consistent with the constitutional deference paid to provinces on most natural resource concerns. Canadian federal government involvement in this policy area has largely been confined to regulation of interprovincial transport of hazardous waste under the Transportation of Dangerous Goods Act.[22] However, the federal presence in hazardous waste policy has gradually expanded since the Canadian Council of Resource and Environmental Ministers began to examine the issue in 1987 and the Canadian Environmental Protection Act was enacted in 1988.[23] Both steps are moving Canada toward more uniform definitions of hazardous waste and more routinized procedures in hazardous waste permitting, somewhat similar to the American experience under RCRA and its subsequent amendments, while leaving considerable authority in provincial hands. Canada and its provinces are also making more aggressive use of highly technical environmental impact assessment procedures pioneered in the United States. This action is consistent with the diffusion of some aspects of North American environmental policy at the federal level from Washington to Ottawa (and sometimes to Mexico City), although whether this same pattern occurs among provinces and states has not been well examined.[24]

The regulatory framework imposed by RCRA on the American states is, in keeping with the American policymaking style, more detailed and imposing on its intergovernmental partners at the state level.[25] It is also a stimulus for litigation, although Canada is rapidly catching up in this area, with mounting legal challenges to relevant environmental legislation and a federal judiciary that is increasingly taking an activist stance on some environmental matters.[26] The American legislation provides uniform national standards and permit guidelines for hazardous waste management and has attempted to shift states away from land-based disposal methods. This has contributed to the growing reliance on incineration, a method called into question by growing controversy and a moratorium on further incineration construction imposed by EPA in May 1993. RCRA sets broad themes but nonetheless operates on a delegative, "conjoint" regulatory basis.[27] More than forty states have acquired authority to operate RCRA permitting processes, and nearly all of these states have enacted their own facility siting legislation. Moreover, RCRA does not in any way establish a process for hazardous waste facility siting, leaving this matter almost entirely up to the states. EPA has also been notoriously lax in enforcing the act's provisions or in pursuing an active strategy for monitoring compliance by individual states or waste facilities.

Perhaps the best-known hazardous waste program of either country's government, Superfund, does not directly influence most aspects of state facility siting. Originally intended as a temporary, $5 billion program designed to clean up abandoned hazardous waste dumps, Superfund has evolved into a classically American litigative search-and-destroy mission, consuming billions of public and private dollars but achieving little definitive cleanup. The program has operated on the "polluter pays" principle and can assess huge liability penalties on firms found responsible for prior dumping, even if the relevant amounts were minute or did not violate federal or state law at the time. Such stringent provisions have made "potentially responsible parties" eager to delay expensive settlements in favor of protracted litigation, often involving other dumpers, insurance companies, or respective government agencies. Such delay has added to long-term hazardous waste management challenges, as enormous quantities of waste that may ultimately require movement and treatment remain politically and legally immobile.[28]

Superfund requires individual states to complete "capacity assurance" plans, which are intended to demonstrate that each state has assured hazardous waste disposal, treatment, or recycling capacity for a two-decade period. They must be submitted to EPA if states are to receive Superfund cleanup dollars, which can be considerable. However, the plans are largely perfunctory and are almost universally accepted by EPA.[29] Like RCRA, even this large program does not prevent individual states from designing their own approaches to hazardous waste facility siting and many central aspects of hazardous waste management, much like their Canadian counterparts. Canada has yet to establish such a program at the federal level, although most of the larger provinces have begun to address this problem, borrowing directly in some instances from the American experience.

The Costs to Society of Siting Gridlock

Similar decentralized approaches to facility siting have resulted in common patterns of siting gridlock and attendant policy outcomes in Canada and the United States. Meanwhile, hazardous waste treatment and disposal costs have steadily climbed in both nations and, for at least some regions and industries, adequate capacity remains a serious concern. In the eyes of some observers, the increasingly familiar Nimby phenomenon is reminiscent of the Luddites of early nineteenth century England, when weavers sabotaged efforts to industrialize their trade. For example, the Southern California Waste Management Forum, a government-industry consortium, decried the Nimby syndrome as a "public health problem of the first order. It is a recurring mental illness which continues to infect the public. Organizations which intensify this illness are like the viruses and bacteria which have, over the centuries, caused epidemics such as the plague." [30] Two leading scholarly analysts of the Nimby phenomenon used somewhat softer language but contended that siting resistance stems from a public that badly misunderstands uncertainty and is "neurotically preoccupied with certain specific and rather baroque hazards to physical health." [31] Not surprisingly, many of the strongest critics of the Nimby syndrome have called for policies that would either bulldoze siting opponents into submission or attempt to subdue them with generous side payments.

Such strategies have already been attempted but have almost invariably failed to secure siting agreements when employed in recent years. They also reflect a limited understanding of what motivates Nimby activists. In most cases, Nimby represents a realistic response to siting processes and to involved governmental and private institutions that give every appearance to communities of being arbitrary and untrustworthy. Moreover, the capacity of Nimby groups to bring siting to a virtual standstill in Canada and the United States since the mid-1980s is not an unalloyed policy failure. The policy ramifications of the Nimby syndrome — and the recent evolution of hazardous waste policy — are mixed.

Public Involvement

Both Canada and the United States have suffered from declining rates of public confidence in public and private institutions. While Nimby politics in certain respects contributes to that lack of trust in prevailing structures, it also illustrates the willingness of the general citizenry to become informed and actively involved on pertinent social issues. The nature of Nimby involvement, in communities of diverse size, economic affluence, and racial composition, demonstrates that political interest and participation may be far greater than surveys and declining election turnout rates might suggest. Rather than view Nimby as a millstone around the neck of regulatory efficiency, it can be viewed as an opportunity to move toward a more open, effective environmental regulatory system in North America. Such broad involvement may prove particularly crucial given that so many sources of hazardous waste exist on the continent and that any long-term solutions may require a broad public-private partnership instead of a narrow, top-down strategy.

Improved Safety

The practice of waste management began to undergo a transformation around 1980, with increasing reliance on more sophisticated types of landfills, incineration, and underground injection. As recently as the late 1970s approximately one-half of all hazardous waste generated in the United States was disposed of in unlined surface impoundments and lagoons, with the remainder deposited either in barrels, solid waste landfills, or municipal sewers or dumped indiscriminately on land or in water.[32] Congress estimated that at least 30 million to 35 million tons of hazardous waste were dumped on the ground each year. Waste horror stories abounded. In New Jersey during this period, for example, warehouses secretly loaded with hazardous wastes exploded, toxic-triggered fires raged for months beneath urban areas, and trucks routinely released wastes during their highway routes.[33]

Canada had similar experiences into the early 1980s, although it relied even more than the United States on dangerous disposal methods and was slower to stop them. In both nations, many of these practices have come back to haunt communities, corporations, and agencies, as these wastes often pose major contamination problems that necessitate complex cleanup in later decades.

Citizen group pressures and expanded regulatory oversight have helped eliminate some of the most egregious disposal practices. They have also intensified pressure on waste generators to either recycle their wastes or alter their practices to reduce or eliminate waste in the first place. Reliable measures of the volumes of wastes being recycled or minimized do not exist for either nation, as federal, provincial, and state governments are only beginning to devise methodologies to make such calculations. But long-standing concerns over the adequacy of treatment and disposal capacity have been eased, at least in part, by the unexpected growth — and relative inexpense, in many instances — of waste recycling and reduction.[34]

Technology

Other approaches to waste management are available, aside from the dominant ones of incineration and landfilling, as well as the slipshod practices of the past. The array of new technologies include distillation, neutralization,

ion exchange, and reverse osmosis. They are being utilized with growing frequency outside North America, particularly in Europe.[35] Many of these offer efficient, safe alternatives to prevailing disposal methods. However, they are used only haltingly in the United States and Canada, in part because of Nimby resistance to new facilities of any sort.

Annual surveys of hazardous waste management in the two nations confirm the extreme difficulty of bringing on line any new disposal or treatment facilities, regardless of the technology proposed. Most capacity that is added represents incremental expansion of extant facilities — and technologies — not bold new steps.[36] As a result, most hazardous waste generators continue to rely on more traditional disposal technologies, either off-site or, more frequently, on-site.

However, blame for the inability to utilize promising new technologies cannot be placed solely with Nimby resistance to siting. The most prominent firms in waste management, including Laidlaw in Canada and WMX Technologies (formerly Waste Management), IT Corporation, and Browning-Ferris Industries in the United States, among select others, have hardly been aggressive and effective advocates for the introduction of new technologies with industries, governments, or the general public. These firms are widely thought to hold a dominant share in most off-site waste disposal markets, ones that are generally difficult for smaller entrepreneurs to enter because of high technology development and potential liability costs. Their reputation is further tarnished by the considerable role that organized crime is widely alleged to play in large-scale hazardous waste management in the United States and, to a lesser extent, Canada.[37]

Not only have the prominent companies been repeatedly mired in regulatory controversies, ranging from lax incinerator oversight to cartel-like price fixing, but they have also continued to emphasize prevailing and profitable technologies of incineration and landfill above all else.[38] Moreover, they have played an active and effective role in lobbying against efforts to accelerate the phase-out of these traditional technologies.[39] In the words of one prominent industry observer: "I don't know many people outside the hazardous waste business who feel it is dynamic. That is a self perception which is erroneous. . . . I think it is a business that is encrusted and it's got some big players who want to keep it that way." [40] Industry leaders naturally refute such observations, citing shifting government regulations that prompted them to invest heavily in technologies — such as incineration in the early 1980s — that have since fallen out of favor. They also note declining profit margins and, in some instances, facility utilization rates, as examples of their plight. Nonetheless, the intensifying debate over the willingness and capacity of waste management firms to play a lead role in developing and operating new technology calls into question the traditional practice of solely blaming Nimby groups for any delay in the introduction of new technologies.

On-Site Emphasis

The growing cost of incineration or landfill disposal has prompted many hazardous waste generators in Canada and the United States to seek on-site solutions to waste management. More than 90 percent of the hazardous waste generated in the two nations is disposed of in this manner; for example, EPA estimated that 95.4 percent of hazardous waste generated in the United States in 1989 by so-called large-quantity generators was disposed of on-site.[41] If anything, this percentage may have increased in recent years as generators prove more reluctant to ship wastes to relatively expensive facilities that use controversial technologies and are under growing public scrutiny.

Remarkably, Nimby groups, other environmental organizations, and governmental agencies have not focused much attention on these facilities, even though they handle a wide array of hazardous wastes and often operate in heavily populated areas. In one sense, these de facto disposal facilities offer a significant advantage over alternative methods in that no risk related to waste transport to an off-site facility arises. However, the little that is known about them is cause for concern. First and foremost, on-site disposal occurs in the virtual absence of federal, state, or provincial regulatory oversight in the United States and Canada. RCRA provides an "interim status" category for such facilities but allows them considerable latitude in selecting disposal technologies and calls for an absolute minimum of direct monitoring. As environmental policy analysts Daniel A. Mazmanian and David L. Morell noted, on-site hazardous waste in the United States is typically "placed as liquids into unlined pits, ponds, and lagoons or other surface impoundments on the production site. No cradle-to-grave manifesto exist for these wastes, some 264 million tons annually." [42] Underground — or deep-well — injection is also a common disposal method, often leading to groundwater contamination.

No comparable set of federal guidelines has been established in Canada, and most provinces have proven more timid in addressing this issue than their state counterparts. As in the United States, on-site disposal has drawn little political or regulatory attention, and this implicitly encourages extended use of highly suspect waste management methods. The issue may be on the verge of drawing added attention, however, at least in the United States. In May

1993 EPA imposed a series of new controls on hazardous waste incineration, a move prompted in large part by concerns over the safety of 171 industrial furnaces that incinerate hazardous wastes and operate under RCRA interim status.[43] The suspect nature of on-site waste management and the massive quantities being processed in this fashion are among the most compelling reasons for development of new, more sophisticated systems.

Long-Distance Transport

Wastes that cannot be treated or disposed of at the site of their generation must be shipped elsewhere. As of 1989, twenty-four states lacked any off-site incinerator capacity; thirty-five states lacked any off-site landfill, land treatment, or deep injection well capacity; eighteen states lacked any off-site biological, physical, or chemical treatment capacity; and twenty states lacked any off-site solvent recovery capacity.[44] Similar gaps existed across Canada and its provinces. Questions emerge concerning the safety of the final handling of wastes and the risks imposed by their long-distance transport to a final destination. As the overall number of facilities has declined in favor of larger, more sophisticated ones, and few facilities can be opened because of Nimby opposition to siting, the issue of lengthy transport has become an increasing matter of interest. And given that so few communities — or states and provinces — are in any way prepared to address hazardous waste spills with existing public safety resources, a growing number of them have attempted to restrict any movement of wastes within their boundaries, whether by truck, rail, or other transportation mode.

The U.S. Department of Transportation estimates that one-half million off-site movements of hazardous waste occur each day in the United States. A truck carrying hazardous cargo, such as wastes, passes a point on Chicago's Dan Ryan Expressway every 22 seconds.[45] Observers confirm similar patterns in Canada, adjusting for differences in population and waste volume.[46] Both nations have experienced disasters in hazardous waste transport. In Ontario, a train derailed in the town of Mississauga, a Toronto suburb, in 1979 and threatened to release chlorine gas. This prompted the evacuation of 240,000 residents and triggered a series of long-term concerns over public exposure to toxic materials. In the next two years, 106 spills of "dangerous goods" from rail accidents and 175 spills from trucking accidents were reported. The Transport of Dangerous Goods Branch of Transport Canada later acknowledged that the figures were "grossly underestimated." [47] Serious spills have persisted in Canada, although none has approached the scope of the Mississauga episode. A riveting example involved polychlorinated biphenyl (PCB)-contaminated waste that completed a transoceanic voyage before being returned to its province of origin, Quebec. Wastes that had been illegally stored in a warehouse in St.-Basile-le-Grande, forty kilometers southeast of Montreal, caught fire in 1988, resulting in the evacuation of 3,000 residents. The residuals from the fire were highly toxic, and Quebec lacked facilities for treatment or disposal. After efforts failed to transfer the waste to Alberta, officials shipped it to England for incineration at a facility in Wales. But after British dockworkers refused to unload it and related protests ensued, Quebec agreed to bring the waste back home, where it remains in a storage facility.[48] Such incidents are consistent with those in the United States, as hazardous wastes and other hazardous materials have been released in accidents leading, in many instances, to loss of life and extensive property damage.[49]

Controversy surrounding long-distance shipment of wastes is exacerbated when jurisdictional boundaries must be crossed. Alongside the more common practice of interstate and interprovincial waste shipping, some hazardous waste crosses the national boundaries of North America. Approximately 5,000 border crossings involving hazardous waste occur each year at the American-Canadian or American-Mexican borders, although this is a highly imprecise estimate, given the paucity of monitoring and enforcement, and in all likelihood underestimates total shipments.[50] No reliable estimate can be made on the volume of hazardous waste that passes the American-Mexican border, despite formal protocols on the matter and the growing attention drawn to this issue during the approval of the North American Free Trade Agreement.[51] Most observers concur that American-Canadian waste trading has increased, with particularly strong growth in recent years from Canada (primarily Ontario) to American waste treatment and disposal facilities, as indicated in table 1-1. Much of this export growth stems from Ontario's profound difficulties in facility siting, which has prompted provincial waste generators to look for long-distance options, particularly in Quebec and American states. In 1992, 8.6 percent of the hazardous waste generated in Ontario was shipped to the United States for final handling.[52] In turn, a waste disposal facility near Montreal has accepted more than 250,000 tons of American hazardous waste since 1985, involving long-distance transport from states such as New Jersey and Massachusetts that have been unable to develop their own treatment and disposal capacity because of Nimby pressures.

Questions arise about the environmental and public health desirability of long-distance waste transport as well as the issue of intergovernmental equity. In the United States, states such as Alabama, Ohio, and South Carolina devel-

Table 1-1. *U.S.–Canada Hazardous Waste Trade, 1987–90*
Tons

Waste flow	1987	1988	1989	1990
Total hazardous waste exports from United States to Canada	130,000	145,000	150,000	143,000
Total hazardous waste exports from Canada to United States	45,000	66,000	100,000	137,000
Net U.S.–Canada waste flow	85,000	79,000	50,000	6,000

Source: Adapted from David Stamps, "Hazardous Waste Exports," *EI Digest: Industrial and Hazardous Waste Management* (July 1991), p. 120.

oped extensive waste treatment and disposal capacity in the 1960s and 1970s, before serious controversy emerged over hazardous waste management or facility siting. As Nimby opposition has mounted, other states, such as Massachusetts, New Jersey, and North Carolina, have been largely unsuccessful in developing anything approaching comparable treatment and disposal capacity. As these earlier, often massive facilities continue to operate, import states increasingly contend that they have become magnets for wastes generated by other states that enjoy all the economic benefits related to industrialization but encounter few of the direct pains associated with waste management.

Many of the states that consistently rank highest in environmental regulatory rigor or effectiveness tend to be least able to develop adequate treatment and disposal capacity and thereby become most dependent on long-term shipment to other states or nations. In North Carolina, for example, which generally receives high marks in surveys on state commitment to environmental protection, particularly in comparison with neighboring states, 81 percent of its 91,314 tons of hazardous waste to be managed off-site was exported in 1990. The majority of this waste was shipped to Alabama, Louisiana, and South Carolina, but significant quantities were also directed to Georgia, Indiana, New Jersey, Pennsylvania, Tennessee, and Virginia.[53] Existing intergovernmental tensions were exacerbated, and, as a result, South Carolina governor Carroll Campbell requested that federal officials bring pressure upon North Carolina to minimize future waste exports. Such tensions are increasingly evident across state, provincial, and national boundaries in the United States and Canada. They have been fostered by Nimby opposition to siting, and the issue of export control has become central to the prospects for future siting agreements.

Racial and Class Equity

The controversy over equitable distribution of the burden of waste management is not confined to governmental boundaries. Mounting evidence indicates that the current distribution of waste treatment and disposal activity is disproportionately concentrated in communities that have large numbers of ethnic and racial minorities and families with low incomes. Sociologist Robert D. Bullard, speaking for a growing body of scholars in this area, contended that "government and private industry in general have followed the 'path of least resistance' in addressing externalities" such as waste facilities and pollution more generally.[54] In turn, he argued, "Nimby has operated to insulate many white communities from the localized environmental impacts" posed by proximity to various waste treatment and disposal facilities.[55]

This issue has risen rapidly as part of the conflict over hazardous waste management, especially in the United States, and further complicates the issue of facility siting. On the one hand, past evidence suggests that market-driven siting strategies of the 1960s and 1970s made siting decisions on the basis of weakest political resistance rather than optimal technical criteria. As environmental policy analyst Michael R. Greenberg and colleagues noted in their analysis of hazardous waste facility siting practices, "The most consistent social effect was that sites tended to be located or potentially located in areas with many relatively poor people." [56] Such host communities, or potential host communities, have become sensitized to this issue and increasingly use equity arguments to seek closure of extant facilities or reject newly proposed ones. On the other hand, equity concerns suggest that any new facilities should be located in more affluent communities. However, these communities can be expected to deflect siting proposals with great success, given their economic and political resources, at least as long as prevailing approaches to siting continue.

The equity issue is further compounded by the increasing practice of waste management firms to target Native North American lands as prospective facility sites. Firms find these attractive both because of their exemption from select governmental regulations and a general expectation that desperation for economic diversification will likely overcome environmental objections. Such siting negotiations are currently under way in places such as Arizona, California, Florida, Mississippi, Nevada, New York, Ontario, and Utah.[57] The Nimby response is increasingly evident in many cases involving Indian or aboriginal lands and further underscores the larger question of creating a politically and socially equitable approach to hazardous waste facility siting and waste management.

Signs of Convergence

The Canadian and American cases of hazardous waste facility siting and hazardous waste management are striking for their many similarities. The cultural and institutional differences in environmental policymaking in these neighboring nations are significant, and some contribute to the finding that siting agreements are somewhat easier to reach in Canada than in the United States. Nonetheless, a central theme of this book remains that a tremendous complementarity exists among these cases, with remarkably similar political scenarios that lead to siting gridlock. In many respects, these cases are integrated, not only because of comparable policy styles and outcomes, but also because of the high amount of technology, information, regulatory techniques, and hazardous waste that moves frequently and freely across various state, provincial, and national boundaries. Canada and the United States, for better or worse, have created a common political and economic market in hazardous waste management.

The bulk of comparative environmental policy analysis focuses on national institutions and processes and accentuates significant differences. It is particularly well developed in comparing the command-and-control regulatory system in the United States with that in its somewhat less adversarial Western European counterparts, such as France, Germany, Sweden, and the United Kingdom.[58] By contrast, the comparative analysis of environmental policy between the United States and Canada is not nearly as well developed. Important work has been completed on the varieties of agenda setting at the federal level and the respective roles of interest groups in the regulatory process.[59] Generally, however, this work has shied away from subnational (state and provincial) comparison, careful empirical examination of specific environmental policy areas, or consideration of convergence of regulatory styles or outcomes.

Moving from the federal level to the subfederal level in the United States and Canada, greater differences may be found among provinces or states within the same nation than among separate provinces and states. Ontario, for example, has designed an approach to siting that bears far greater resemblance to that in American states such as New Jersey, New York, and North

Carolina than that of any other province. In turn, other provinces such as British Columbia and Saskatchewan have developed approaches highly analogous to those employed in other American states. Nonetheless, significant differences endure and must be recognized.

In some respects, Canada could prove more hospitable to proposed sites than the United States because of its distinct style of governance and political culture. However, the influence of these factors in fostering siting agreements must not be exaggerated, especially given the frequency and similarity of the Nimby reaction to siting proposals in so many provinces and states. Canada has hardly emerged as an easy mark for siting agreements. Instead, governments and waste management firms on both sides of the forty-ninth parallel continue to struggle in seeking such agreements. The most significant differences include the following.

Political Culture

Many scholars have emphasized the more consensual, less adversarial political culture of Canada.[60] Political sociologist Seymour Martin Lipset spoke for a wide range of analysts when he observed that Canada "remains more respectful of authority, more willing to use the state, and more supportive of a group basis of rights than its neighbors." [61] Applied to hazardous waste policy, this might generate a calmer, more workable approach to siting in most, or perhaps all, provinces, as diverse constituencies could work harmoniously in pursuit of the common good of effective waste management. However, in every province in which serious efforts have been made to site facilities, Nimby-type reactions have emerged to block proposals. Even the provinces that have had some success in reaching a siting agreement did so only after earlier failures that closely resembled the American, adversarial style. In provinces such as British Columbia, Ontario, and Saskatchewan, Nimby continues to prevail, transcending the capacity of a more consensual political culture to dampen conflict when an issue as volatile as hazardous waste facility siting is under consideration. And in Quebec, such conflict has reemerged as the province has failed to follow the promising approach toward hazardous waste that it developed in the early 1980s. Nimby conflict is also pervasive in other areas of waste management across Canada. In many provinces, for example, the inability to site new solid waste facilities has led to far greater concern over garbage crises than in the United States.

If anything, Canadian environmental policy is moving in the more litigious, adversarial direction of American policy, dampening any effects of its distinctive political culture.[62] This can be attributed to the increasingly expansive judicial role being assumed by the Supreme Court of Canada in environmental regulatory matters, the proliferation of environmental advocacy groups and their adoption of American-style strategies, and growing public commitment to aggressive environmental protection as reflected in public opinion surveys.[63] More generally, Canadian politics and public policy may be following the contentious American pattern, particularly given the numerous fissures in the Canadian citizenry evident in the 1992 referendum rejection of a proposed constitutional accord, the 1993 federal elections in which regionally based parties emerged as central political forces, and plummeting public confidence levels in political leaders and institutions.[64]

Unified Government

Canada and its provinces are distinct from their American counterparts in that their reliance on a parliamentary system of governance fosters greater party unity and executive control.[65] During the period of greatest controversy over facility siting in the United States, the federal and many state governments have experienced "divided government," whereby both parties formally share executive and legislative powers. In the eyes of many scholars, this has compounded the problems of devising and implementing coherent policy.[66] By contrast, the more unified, executive-driven Canadian provincial governments could be more able to devise a manageable, orderly approach to siting.[67] As political scientist Milton J. Esman noted, Canadian "provincial premiers have the power to organize the executive units of their governments, subject them to programmatic discipline, and thus gain control over program specialists." [68]

In practice, however, how this affords any significant siting advantages to Canadian provinces is not clear. For example, the case of Ontario suggests that despite strong, consistent provincial direction from the Ministry of Environment and the Ontario Waste Management Corporation, public distrust is rampant and has resulted in a twelve-year standoff in hazardous waste facility siting in the province. Much like the experience of American states such as Florida, New Jersey, and New York, subfederal government efforts to impose facilities on communities seem unlikely to result in siting agreements, regardless of party unity or executive branch centralization. The American cases do not indicate any difference in style or outcome of siting policy whether the state government is unified or divided by party.[69]

Federalism and Regulatory Style

Constitutional and political realities delegate greater environmental regulatory power to Canadian provinces than American states. Canada has no full counterparts to American programs such as Superfund and RCRA. These American programs add considerable complexity to the process of hazardous waste management and provide targets for prospective litigators and political combatants. Compliance with federal permitting requirements is more exacting and laborious in the United States, as is any exercise in environmental impact assessment.[70] Moreover, EPA assumes a more adversarial posture toward states than does Environment Canada toward provinces, suggesting less likelihood of intergovernmental flexibility.

Although significant, these differences should not obscure the fact that in both nations considerable authority in hazardous waste facility siting and management is delegated to provinces or states. Provinces are free to go beyond the basic framework of the Canadian Environmental Protection Act; states are free to go beyond the basic framework of RCRA. As policy analyst Mario Ristoratore, one of the few scholars to conduct comparative analysis in this area, noted, "Federal governments in both Canada and the United States have no jurisdiction on the siting of non-nuclear waste facilities." [71] Examination of the respective cases from both nations suggests that provinces and states take considerable advantage of this latitude and encounter few federally imposed restrictions on their siting efforts. Even the requirements for states to secure twenty-year capacity assurance plans under Superfund are gingerly enforced and have not resulted in a single financial penalty on states. Neither RCRA nor Superfund would preclude states from taking the voluntary approach to siting, employed so successfully in Alberta and Manitoba, should they choose. Individual states have considerable freedom in adapting some of these characteristics for their own siting processes.

Cross-national distinctions of federalism and regulatory style diminish further, at least in hazardous waste policy, when the design and implementation of policy are considered. The distinctions may be exaggerated in much of the comparative literature because of the limited systematic comparison of American states and Canadian provinces. Instead, much of the literature on comparative federalism and intergovernmental regulatory policy tends to reflect a top-down approach that accentuates broad, formal differences but lacks the systematic, comparative case design necessary to examine policy differences at the state and provincial levels. Moreover, much of it has a detached, armchair quality that suggests minimal field work or direct examination of central questions. As a result, much as Ristoratore concluded in the mid-1980s, "many other generalizations about U.S.-Canadian differences . . . do not always apply at the state-provincial level." [72]

Factors Fostering Convergence

At least three factors foster greater convergence across states and provinces than might be anticipated. First, the steady export of hazardous waste across national boundaries creates a de facto, binational waste management market. As noted in table 1-1, waste trading between the two nations is an increasingly common occurrence. For some states, such as Connecticut, Massachusetts, Michigan, Minnesota, and New Jersey, access to Canadian waste treatment and disposal facilities is an essential component of their waste management strategies. Provinces such as Alberta (before the opening of its comprehensive facility in 1987), British Columbia, Manitoba, Ontario, and Saskatchewan have long been reliant on American-based treatment and disposal facilities as part of their waste management strategies. For both nations, conversance with waste regulations across borders has become increasingly important to their basic functioning. The diffusion of environmental regulatory ideas across provincial and state boundaries may be far greater than generally realized. Ontario consistently uses American states as a model for its environmental regulatory reform efforts and in many respects has followed the pattern of neighboring American states in designing its approach to siting and waste management.[73] A number of states have begun to turn to successful Canadian siting models in attempting to redesign their own policies, both in hazardous and low-level radioactive wastes.

Second, the practice of waste management in Canada and the United States remains dominated by large corporations, many of which operate in both nations. A tradition also exists of firms based in one nation owning or holding a large stake in for-profit facilities located in another. For example, an American firm owns and operates Quebec's largest waste disposal facility, while a Canadian firm owns and operates one of the largest medical waste incinerators in the United States, located in South Carolina.[74] The leading industry lobbying group, the Hazardous Waste Treatment Council, represents clients from both nations. Many Canadian facilities that accept American wastes have formed the International Environmental Policy Coalition and have retained an influential, Washington-based law firm to lobby against any pending congressional efforts to tighten controls on American waste exports to Canada.[75]

These efforts have been highly successful in maintaining open boundaries and creating a binational waste market. Although RCRA requires that American waste exports receive a "consent notification" from an importing government, a 1986 bilateral agreement actively sought by Canada assured that this would be waived. As a result, waste shipments between the two nations proceed without any formal consent by the importing government. They are stopped only when a formal objection is raised, which rarely occurs. Consequently, the American and Canadian waste management markets have in effect merged into one.[76]

Third, the legal and technical languages used to define hazardous waste and set treatment and disposal priorities are analogous across state and provincial borders. This may reflect, at least in part, a historic continuity. In an analysis of comparative regulatory regimes, policy analyst Fred Thompson noted that "until very recently, American and Canadian regulatory regimes were remarkably similar: both countries regulated many of the same things and in much the same ways. Furthermore, despite differences in political institutions and structures, the regulatory policies and practices of the two countries produced very similar patterns of winners and losers." [77] In the same volume, policy analyst Thomas L. Ilgen highlighted important differences in federal policy that have developed for various types of toxic substances.[78] But since then, the more historic convergence has begun to reappear, especially when moving beyond the details of federal legislation and regulatory style to the state and provincial levels, where the crucial decisions on siting are made. In other areas of domestic policy, such as those related to economic development, the diffusion of policy ideas and development of comparable, cooperative strategies across state and provincial borders are commonplace.[79]

These comparative distinctions are important and may pose impediments to any effort by one nation to base reform on the experience of its neighbor. Nonetheless, the signs of case convergence are significant, pointing to surprisingly similar regulatory approaches to facility siting, and for the most part, strikingly similar Nimby outcomes.

2

When Siting Does Not Work

THE DOMINANT APPROACHES to hazardous waste facility siting in Canada and the United States have given communities and individuals little reason to do anything other than attempt to block facilities once they have been proposed. Whether advanced by private or public entities, siting proposals are characteristically made with no prior consultation with targeted communities. They are neither linked to any long-term strategy for waste management for a province or state nor advanced with any guarantees of protections against acceptance of certain types of wastes or wastes generated in other parts of the continent. Instead, siting is often presented in an arbitrary, ad hoc manner — a take-it-or-leave-it proposition that community after community has chosen to reject.

Before the late 1970s siting was far less complicated or conflictual. Private facility developers were encouraged by governmental agencies such as EPA to maintain low profiles and minimize community contact. Calming words, including *recycling* and *industry support services*, were used in naming the facilities; little, if any, public notification of new facilities was made before their opening; and state and provincial agencies were happy to speed the facilities through the modest permit procedures in existence at the time. Many facilities were located in low-income, predominantly racial minority communities, those most desperate for economic development and least likely or able to make a political fuss. Through such low-key operations, much of the existing waste management infrastructure of landfills and incinerators was put into place, including the massive facilities that continue to operate in, for example, Alabama and Ohio. Many functioned with an absolute minimum of technical safeguards or provisions for community input or oversight of facility management.

The 1980s brought a halt to this quiet practice in both Canada and the United States. A series of episodes, including the discovery of extensive toxic contamination at Love Canal in New York, drew unprecedented attention and pushed the issue of facility siting high onto national, state, and provincial environmental agendas. As a result, private waste management firms experienced unanticipated public opposition to proposed facilities, as virtually every state and province encountered some form of Nimby reaction once siting plans were revealed. In response, most states and provinces responded with efforts to formalize siting processes. These commonly added new regulatory layers to traditional permit processes and created distinct bureaucratic settings for overseeing siting and waste management. New federal legislation had some impact, particularly in the United States, but left enormous decisionmaking latitude to individual provinces and states.

The varying approaches that have been adopted by provinces and states are fundamentally divided by who makes siting decisions: governmental agencies, volunteer communities, or private waste management firms. To examine the siting process, I selected at least two cases in each nation for each siting venue. In some instances, venue changes occurred during the course of the study (see figure 2-1).

Provinces such as Ontario and states such as Arizona, Florida, Georgia, New Jersey, New York, and North Carolina have relied upon provincial or state environmental or natural resource agencies to make the main siting decisions and impose them on local communities (cell 2). British Columbia has also experimented with a variant of this approach. Under these "regulatory" approaches, governmental officials weigh a number of siting criteria — technical, economic, social, and political — and decide what type of facility is necessary and where it should be located. Local governments and the general public are usually not consulted until after the criteria have been applied and one or more preferred sites have been decided upon. A variety of coercive or consensus-seeking methods may then be used to either force construction of the new facility or gain local support for it. Private corporations with expertise in various aspects of waste management may be included on a contractual

Figure 2-1. *Approaches to Hazardous Waste Facility Siting Policy in Canada and the United States*[a]

	Centralized process	Decentralized process	
Regulatory approach	Canada HLRW U.S. HLRW	Arizona[b] Connecticut LLRW Florida[b] Georgia Michigan LLRW New Jersey New York North Carolina[b]	British Columbia[b] Ontario
	1		2
Voluntary approach	Canada LLRW	California Greensboro, N.C. Minnesota Nebraska LLRW	Alberta Manitoba
	3		4
Market approach		Arizona[b] Florida[b] Illinois Massachusetts Michigan North Carolina[b]	British Columbia[b] Maritimes Quebec Saskatchewan
	5		6

a. In a centralized siting process, the national government is dominant; in a decentralized process, the state or provincial government is. In a regulatory approach, one or more government agencies make the main siting decisions. In a voluntary approach, siting proceeds only among communities that are willingly participating. In a market approach, private sector developers make the main siting decisions. HLRW = high-level radioactive waste. LLRW = low-level radioactive waste.

b. Employed multiple strategies either simultaneously or at different periods during the course of this study.

basis. They may be hired to construct and operate the facility after the provincial or state officials have determined its location, the wastes that will be accepted, and the methods that will be used for their disposal and treatment.

Quebec, Saskatchewan, and the four Maritime provinces and states such as Illinois, Massachusetts, and Michigan give their public officials a far more passive role in the siting process (cell 6). Arizona, British Columbia, Florida, and North Carolina have also experimented with this approach, to varying degrees, as did Alberta and Manitoba before undertaking fundamental reforms. Private sector initiative drives the siting process under this "market" approach. After establishing general guidelines for safety and facility operation, provinces or states wait to receive proposals from private facility developers to specify the site, the types of wastes to be accepted, and the nature of the facilities to be constructed. These private developers attempt to secure support from communities that they have selected as candidates to host a site. They may interact directly with the proposed communities or through neutral mediation services provided by the government. In the absence of proposals from the private sector, no new facilities will be developed.

Both provinces and states have devised remarkably similar siting programs, with most choosing either a regulatory or market strategy (or some combination). And in both nations, these respective approaches have consistently failed to produce agreements on hazardous waste facility siting. Among regulatory approaches, even the existence of a dominant governmental authority is insufficient to overcome local resistance.[1] Instead, enormous public distrust often is triggered by any governmental role in siting. Among market approaches, the attempt to establish a workable bargaining process in the absence of a substantial governmental role faces similarly rigid public resistance. Time and again in states and provinces, private site proponents withdraw their proposals in response to fierce local outcry.

In contrast, those cases in which volunteer communities, not government agencies or private firms, make their own siting decisions, appear far more likely to result in siting agreements. The Alberta, Manitoba, California, and Minnesota cases, which took the "voluntary" approach, are clustered in cell 4, and a case from Greensboro, North Carolina, can also be placed there. Quebec's successful siting effort of the early 1980s in many respects warrants placement in cell 4, although much of its later siting activity falls within cell 6. Canada's approach to low-level radioactive waste (LLRW) facility siting is placed in cell 3, reflecting its volunteer characteristics, while Nebraska's LLRW siting effort mostly belongs in cell 4. Other factors, such as economic compensation, are not irrelevant to eventual siting agreements, but the siting decision remains with the individual community and siting can proceed only if such volunteers emerge.

The typology in figure 2-1 also accounts for differences in degree of centralization in siting. All American and Canadian hazardous waste cases are placed in the "decentralized" cells (2, 4, and 6) because of the substantial role delegated to provinces and states. However, the Canadian cases are placed at the far end of this continuum given the lesser involvement of its federal government in hazardous waste policy. The complementary cases of radioactive waste facility siting in the two nations are located elsewhere along this continuum. Both nations have taken centralized regulatory approaches to high-level radioactive waste facility siting, with little success. For low-level wastes, the United States has adopted a somewhat decentralized approach involving multistate compacts under a tightly prescribed federal regulatory framework, generally with little success. Canada, meanwhile, has devised an approach to LLRW facility siting that emulates many of the features of the fruitful hazardous waste siting cases. The American case that has come closest to a siting agreement, host state Nebraska in the Central Compact (subsequently renamed the Central Interstate Compact), has followed a similar pattern. What appears most crucial to elevating the chances of siting agreements is not the degree of decentralization but moving beyond the regulatory and market approaches to siting still so dominant in both Canada and the United States toward the voluntary approach that gives considerable signs of promise.

As practiced, the prevailing approaches to siting boil down the entire process to a head-to-head confrontation between site proponent and designated site host.[2] Under the market and regulatory siting approaches, site proponents have strikingly similar incentives. In market-based siting, private site developers want to construct and operate a facility with minimum delay and expense, so that they may maximize profits in handling hazardous waste. In regulatory-based siting, government agency officials have a strong desire to open a facility quickly and thereby assure waste-generating entities (many of which contribute significantly to state or provincial employment and tax revenue bases) a reliable method for waste management. Groups opposing site proposals may threaten either corporate or governmental preferences and may not appear to be reliable partners in any broader discussion of waste management and burden-sharing options. Not surprisingly, private and public site developers refrain from consensus-building procedures in an effort to override any local objections. This helps explain the common state and provincial siting pattern that eschews prior consultation in an effort to push through a specific siting proposal as swiftly as possible, whether through market enticements such as compensation or regulatory powers such as preemption.

In response, citizens and their local community organizations may view potential site developers with similar disdain. Often, they have never thought about waste management, much less the possibility of hosting a major waste treatment or disposal facility, before learning they have been selected by a corporation or agency as the preferred candidate.[3] In the words of one Massachusetts resident, after learning of his community's selection for siting, "It's a shock to be named. Nobody ever thinks his community will be singled out." [4] As a result, citizens may perceive their solitary goal as termination of further siting deliberations as quickly as possible, thereby minimizing any possible exposure to hazardous wastes that might threaten their health or property values. They can be expected to take whatever political action is necessary to eliminate this threat, even if it indirectly involves transferring responsibility for ultimate waste management to some other community, province, state, or nation. In so doing, these individuals and their groups will have a variety of possible political methods that can be used to scuttle site proposals. They will face few problems in taking collective action, given the strong incentives to participate, and may utilize adversarial approaches, such as protest and litigation, to stall or ultimately reject specific proposals.

Under such circumstances, highly conflictual processes of dispute resolution will prevail, as they generally have in hazardous waste facility siting in both Canada and the United States in recent years. Although some mutually accepted compromises have been reached — for example, through hefty compensation packages to economically desperate local communities — they have done so only rarely since 1980. Instead, conflict has prevailed, as both sides fail to develop cooperative approaches that would result in a mutually satisfactory outcome and some broad agreement on hazardous waste management.

In Canada and the United States, local communities retain the upper hand in strategic siting interaction. They are able to reject siting proposals with remarkable effectiveness and speed, regardless of whether market or regulatory approaches are employed. The most common outcome is that the potential host community takes such a hostile position that the site proponent deems further consideration pointless and withdraws its proposal. Upon withdrawal, the community usually celebrates its having thwarted the site proposal. But the repeated blockage of specific sites does not resolve the problem of hazardous waste management . To the contrary, it creates a host of new environmental and political problems.

Market Approaches: The Limits of Compensation

Persistent failure to reach siting agreements has not deterred many states and provinces from continuing to adhere to market-based approaches to hazardous waste facility siting. These strategies tend to be modified with sufficient frequency that precise categorization at a given time is difficult, although recent surveys suggest that at least twenty-eight states and seven provinces rely exclusively on a market orientation in their siting efforts. Moreover, approximately ten other states and the remaining three provinces allow for some form of a market approach as a supplement to other strategies.[5]

The market approach offers considerable attractions to state and provincial governments. First, it confines a government's role in hazardous waste to regulatory oversight. Private corporations must bear the economic costs — and political risks — of selecting sites, securing agreements, and operating facilities if successful. This minimizes direct costs to the public sector and also means that governmental officials do not have to assume responsibility for defending controversial site selection. Second, this approach retains the traditional private sector responsibility for waste management long established in Canada and the United States. Hazardous waste is treated as an inevitable by-product of commercial activity, one for which private management solutions are preferable. Third, the market approach allows large private waste management corporations to dip into their ample financial reserves and test the theory that economic compensation can assuage community opposition to facility siting. Most state and provincial strategies of this sort are premised on the assumption that traditional market mechanisms can foster cooperation, with the main business of siting reduced to a negotiation over what price a community will settle for in exchange for hosting a facility. Because the waste management industry is dominated by a relatively small number of large, well-endowed firms, these companies are presumably far more capable of devising extensive compensation packages than any governmental body.

These and related factors have contributed to the enduring popularity of market approaches to siting. Ironically, perhaps the greatest achievement of these approaches is that they have proven so unpopular that they may have blocked a number of facilities that were poorly planned or even unnecessary. The failure of these siting efforts, in turn, sets the stage for a more careful approach to future waste management. But they also offer a series of lessons concerning what should be avoided in efforts to utilize market approaches in any future siting attempts.

Massachusetts

The best-known celebration of policy innovation in Massachusetts in recent decades involves the so-called Massachusetts Miracle, the Bay State effort to bring industrial policy to the state level with unprecedented vigor.[6] The results have been questioned because of the prolonged recession in the state. But a second area in which Massachusetts was recently heralded as a policy pioneer involves hazardous waste. In the late 1970s state officials feared considerable shortages of hazardous waste treatment and disposal capacity and recognized the dismal failure of recent siting efforts that relied exclusively on private firm initiative. In 1980 they responded with new legislation, the Massachusetts Hazardous Waste Facility Siting Act, that was widely perceived in state policy, environmental policy, and academic circles as rich with promise for replacing the increasingly common Nimby syndrome with prolonged negotiation and, ultimately, siting agreements.

Before the new legislation, Massachusetts had relied heavily on out-of-state export for hazardous waste management. The state produces substantial quantities and types of hazardous waste, consistently ranking among the top six states in waste generation. Twelve commercial facilities in the state operate under RCRA interim licenses that allow them to accept some hazardous wastes, and more than fifty firms treat or store their wastes on-site. But of the total volume of wastes that cannot be managed on-site, more than 90 percent must be shipped to other states, provinces, or nations. In 1987, for example, Massachusetts was the nation's largest net waste exporter, sending out 159,197 tons more of hazardous waste than it imported.[7]

This pattern began in the early 1970s but drew little attention until questions arose concerning the viability of export as a long-term waste management strategy. A 1979 announcement by the New York Department of Environmental Conservation that the state would soon stop accepting hazardous waste from outside its boundaries caused great concern because at the time more Massachusetts waste went to New York than any other state. In an initial response, the administration of Massachusetts governor Michael S. Dukakis published a list of eleven possible sites for a major hazardous waste facility.[8] This led to further outcry, particularly in the targeted communities, and resulted in the formation of a nineteen-member commission charged with the creation of a new state policy toward hazardous waste facility siting. The efforts of this commission led directly to the passage of the 1980 siting act by a legislature eager to be seen as doing something constructive about this mounting problem.

The legislation retained private sector initiative in site selection and facility operation but gave state officials some authority beyond that typically established under market siting strategies. The state would remain neutral on site selection and waste management techniques to be employed in proposed sites. Once a private firm announced a preferred site, however, it would have to receive approval of its notice of intent and preliminary project impact report by the Massachusetts Hazardous Waste Facilities Site Safety Council, which had been created by the act. This report was to include a draft environmental impact report as well as a draft socioeconomic impact assessment. The preliminary review process was designed to give potential host communities some reassurances before formal or informal negotiations with the site proponent began.

Massachusetts also attempted to transcend the limitations of traditional market approaches by assuring outlets for early public review of site proposals. Communities selected for potential site hosting received renewable state subsidies of up to $15,000 to form local advisory committees. These committees could use state funds to hire consultants and other expert advisers to assist them in negotiations with firms proposing waste facilities. The state also launched a $375,000 public relations campaign, relying primarily on mass media advertising, to promote the idea of building new hazardous waste treatment and disposal facilities within Massachusetts boundaries.

At the same time Massachusetts sought to formalize community input, it attempted to withdraw traditional legal methods that Massachusetts communities had used to thwart siting efforts in the 1970s. The new legislation provided that waste management facilities could be constructed in any area zoned for industrial use and prohibited communities from amending local zoning by-laws or enacting restrictive land-use permits once a facility had been proposed. It also allowed local boards of health to refuse a site assignment only when it had been determined that a facility posed "special risks." [9]

All of these provisions were intended to buttress the underlying premise of the legislation: that compensation through formalized negotiations between a waste management firm and the local community could result in siting agreements. Once preliminary project impact reports were filed and local advisory committees were formed, direct negotiations were to occur between these parties. And, if an impasse were to occur, the parties were to submit their dispute to binding arbitration.

Much of the strategy developed in the legislation was guided by a growing body of literature and experience that called for a shift from litigative, adversarial dispute resolution procedures to more consensus-building ones. In dozens of cases, a form of "alternative dispute resolution" has led to some resolution of contentious environmental issues.[10] In a different form, this approach has had significant results in other siting cases. But its transformative potential when mandated in the volatile area of hazardous waste facility siting was seriously oversold by academic and other proponents. Instead of ushering in a new era of cooperation in hazardous waste management, the Massachusetts experience has, according to one analyst, created "an even more treacherous climate for facility siting than was already in existence." [11]

In each of the five attempts to site hazardous waste treatment and disposal facilities under the 1980 act, a volatile, classic Nimby reaction ensued that prevented serious negotiations from occurring in any individual case. In Freetown, residents formed a group named Keep Freetown Hazard Free to fight a fairly small inorganic waste disposal plant. A nonbinding referendum in 1983 rejected the proposal by a 1,575-to-471 margin, and the site proponent quickly withdrew its proposal. In Warren, city officials and a group called Stop-IT blocked the efforts of IT Corporation to build a waste disposal facility that, if constructed, would have been the nation's largest and would have drawn extensive wastes from out of state. Even after clearing a series of procedural hurdles and gaining court support for its effort to eliminate legal restrictions imposed by Warren, IT Corporation chose to give up in 1984. In Braintree, state officials were initially optimistic about a proposal to expand incineration capacity at a small, existing waste disposal facility. But fierce local resistance from officials of Braintree and neighboring communities, the local advisory committee, and citizen groups ultimately led to the rejection of the proposal by the state Department of Environmental Protection in 1991. Opposition became so intense that pressure has continued to close the existing facility.

These and other siting cases under the Massachusetts legislation are linked by the common inability of compensation to entice communities to participate. As environmental policy analysts Gary Davis and Mary English noted, the Massachusetts approach "bypasses basic policy questions such as need for the facility and the safety of the technology and goes directly to bargaining over the conditions under which the community would accept a given facility." [12] Jobs and other economic development enticements were dangled before communities. The Braintree case, for example, included proposals by Clean Harbors, the site proponent, to help pay the community's expenses for overseeing the plant in future decades, to purchase for community use fire fighting equipment suitable to contend with chemical fires, and to avoid handling certain kinds of wastes.[13] A few prominent figures in select communities expressed some interest in such compensation packages but were generally drowned out in a strong chorus of opposition. "Even in what should be easy siting cases with very generous compensation packages, communities feel compensation offers will hurt rather than help them," explained a former Massachusetts environmental commissioner, who played an active role in attempting to implement the legislation. "Maybe there are ways to redefine or expand compensation, but I'm not convinced that compensation alone can work in siting negotiations." [14]

The other cornerstones of the Massachusetts approach collapsed because of the refusal of prospective host communities to take compensation seriously and pursue extended negotiations. Some communities defied state provisions against new local land use restrictions; even if these controls were eventually overturned in the courts, they were successful in delaying any serious negotiations. Some communities went far beyond the small amounts of funding provided by state assistance grants, in some instances spending hundreds of thousands of local taxpayer dollars to fight site proposals. These funds were used for a variety of purposes, from examining the technical details of proposed facilities to researching the record of site proponents in other states and nations.

In all of these cases, communities expressed outrage upon learning that they had been selected to bear the burden of hazardous waste management for Massachusetts, and, in the case of larger facilities, much of New England. The state's public relations campaign was confined largely to making a broad argument on the need for new facilities and did not begin its efforts until nearly two years after the legislation was enacted. Its pitch was thus only beginning after siting controversies were already brewing in the communities of Haverhill and Warren. The formation of a Coalition for Safe Waste Management, a consortium of industry, government, labor, academic, and environmental leaders headed by former governor Francis Sargent, was designed to build public support for the process and established a goal of placing a hazardous waste facility in Massachusetts within a year of its creation in late 1981.[15] Like the public relations campaign, however, its efforts were limited and belated, initiated only after individual communities had been targeted for siting in the absence of any larger — or prior — public deliberation or siting rationale.

Massachusetts has devoted more than a decade to experimentation with this approach, with little to show for it. The state continues to score near the very top of virtually all ranking schemes that measure varying degrees of environmental regulatory rigor and innovativeness at the state level.[16] This is reflected, among numerous other possible examples, in the creative way it attempted to structure its market-based approach to siting in 1980 (and in subsequent legislative refinements). But innovative or not, the reality of Massachusetts hazardous waste management in the 1990s remains strikingly similar to that of the early 1980s, a combination of on-site management and out-of-state export that has many undesirable features. Moreover, the Massachusetts emphasis on compensation and negotiation between site proponent and designated host community has only compounded the problems of public distrust of government and waste management firms that contributed to the need for the 1980s reforms, thus making the prospects for any future siting efforts extremely precarious.

British Columbia

The 1980s siting strategy of Canada's westernmost province featured none of the intricacy of the one devised by Massachusetts but seemed a reasonable candidate for success. Its market orientation comported with the general political philosophy of the governing Social Credit party to minimize direct governmental intervention, and many communities were expected to prove receptive to proposals to diversify their economies through hosting a waste facility.[17] In addition, environmental regulations, particularly for hazardous waste, were among the least stringent of any province or state, and no well-established pattern of environmental activism existed in the province. Moreover, British Columbia had a clear need for treatment and disposal facilities. It ranks third behind only Ontario and Quebec in volumes of hazardous waste generated annually and lacks any sophisticated facilities, relying primarily on exports to the United States, on-site disposal, and a series of particularly suspect methods. According to one provincial official, interviewed in 1991, "We believe considerable amounts of hazardous wastes are illegally and inappropriately dumped into the environment." Municipal sewers and solid waste landfills are thought to continue to be leading dumping grounds for provincially generated hazardous wastes.

But much like in Massachusetts, the prospects for siting agreements seem even more limited in the 1990s than in the past decade, given the tremendous controversy surrounding recent provincial siting efforts. These took two distinct forms, both resulting in similar siting failures. British Columbia began to take siting seriously in 1979 through the formation of the Hazardous Waste Advisory Committee, which in 1981 filed a report that called for significant steps, including creation of a public hazardous waste corporation, to begin to foster the trust so lacking in earlier market-style siting efforts in British Columbia. However, the provincial Waste Management Act, passed in November 1982, took a different approach, emphasizing that hazardous waste management must remain an activity dominated by private corporations with necessary technical expertise and fiscal strength.[18] As a result, the province confined its role to soliciting siting proposals from private firms in Canada and the United States.

In September 1983 the British Columbia government chose to endorse the application of a consortium consisting of Genstar Conservation Systems and IT Corporation, the latter being the firm that encountered such profound problems in its siting attempt in Warren, Massachusetts.[19] The government assumed the consortium could reach a siting agreement and operate a comprehensive waste management facility for the province, including landfill, incineration, and various treatment technologies. But this approach collapsed in July 1984, when the consortium withdrew from further involvement. The consortium emphasized "excessively high capital costs" for compliance with provincial treatment standards and questions concerning reliability of waste flows to any facility.[20]

The latter of these two reasons, however, appears to be the most convincing. British Columbia regulatory standards were hardly oppressive in comparison with other provinces and states (including Massachusetts) where these firms remained active and profitable in waste management. Instead, a more serious concern was whether provincial regulations would be enforced with sufficient rigor that waste generators would abandon existing waste disposal strategies and send their wastes to a consortium-developed facility. The consortium estimated that only 15,500 tons of hazardous waste would be available per year, even though officials acknowledge in the 1990s that annual generation estimates of 110,000 tons (with 400,000 tons currently being stored) fall significantly below the actual amounts of waste being generated. A 1982 study concluded that the province generates almost 1 million tons of hazardous waste each year, a figure that has not likely dropped given the relative absence of emphasis on waste reduction in subsequent years.[21] Furthermore, conflict existed over the issue of waste exports, as provincial preference for "zero net flow" in exports and imports reduced the likelihood that the firm could seek large quantities of wastes from outside British Columbia.[22]

The withdrawal of the consortium led the province to modify its hazardous waste facility siting strategy, taking a slightly expanded provincial role in 1987. As before, the province solicited proposals from private firms to construct and operate a comprehensive waste management facility. The province wasted little time in selecting the Envirochem Group, a consortium of four waste management firms: Envirochem Services and Snadwell Swan Wooster of British Columbia, Environmental Systems Company of Arkansas, and Stablex Canada of Quebec.

After this endorsement in January 1988, the province went beyond a purely market emphasis, incorporating some aspects of a regulatory strategy as well as the voluntary approach later used with success in other provinces. Provincial officials sent letters to all British Columbia communities concerning the possibility of hosting the Envirochem Group facility and received written expressions of interest from eight communities. The province then chose to target further siting efforts on three sites in the Ashcroft-Cache Creek area, and its two small, primarily rural, communities located sixty kilometers west of Kamloops. Technical and political considerations were central in this early effort to eliminate other siting options. A nonbinding referendum quickly followed, and in May 1988 about 60 percent of participating voters in the communities of Ashcroft and Cache Creek approved continued discussion. Within weeks, provincial officials approved the site, which would be handed over to the Envirochem Group as it pursued negotiations with the prospective host communities.

The siting process collapsed quickly thereafter, however. Opposition began to build both within the two communities and in surrounding areas. A group named the Hazardous Waste Management Coalition, which was founded in 1980 to block an earlier siting proposal by a private firm, was revitalized and became a leading source of opposition. This organization and similar ones engaged in door-to-door canvassing, initiated a boycott of local merchants who supported the facility, and undertook an antifacility petition drive that garnered a "very large number of signatures from residents." [23] The Hazardous Waste Management Coalition would later claim that 87 percent of residents were opposed to the facility after these activities. Moreover, many individuals from neighboring communities, who were not eligible to participate in the referendum, such as cattle ranchers, farmers, and tourist facility managers, played an active role in opposition. Finally, many community residents were not listed on the voting rosters used for the referendum, potentially excluding a significant slice of the electorate.[24] The first public hearings on the facility were held in November 1988, and after a strong display of public opposition during them, the Ashcroft-Cache Creek area was withdrawn from further siting consideration. The Envirochem Group never actively pursued siting in another British Columbia community.

British Columbia's shift from a pure market approach to one that involved a somewhat larger provincial role in site selection in the 1980s bore some resemblance to the strategies successfully employed elsewhere. However, the voluntary site selection process was extremely superficial compared with the multifaceted public education and participation processes developed in Alberta and Manitoba. "The public was not well informed," explained a senior provincial official. "The Ministry of Environment thought the proponent would do this, and Envirochem thought the Ministry would do this. I guess it was bad planning and bad communication." Both the province and Envirochem were so eager to seize upon communities where elected leaders expressed some possible interest in siting that they moved at an extremely rapid pace and never established the type of prolonged public dialogue so crucial in other siting agreements. As a result, according to another official, "We didn't even get to the point of being able to offer the community any compensations."

Like Massachusetts, these conflicts have left British Columbia hazardous waste facility siting and management in greater turmoil than a decade ago. Even efforts to establish small waste storage and recycling facilities have met with increasing political opposition. In response, British Columbia officials have decided to emulate their more successful Canadian neighbors and form a crown corporation to explore siting alternatives that involve more extensive and systematic public input and waste management options such as recycling and reduction. By the mid-1990s, however, the traditional waste management practices continue to dominate, although long-term reliance on exports may be an unreliable strategy. Some American states are becoming increasingly reluctant to accept British Columbia–generated wastes, and the comprehensive disposal facility in neighboring Alberta that some British Columbia officials and waste generators have eyed as a potential solution to their waste problems has thus far remained unwilling to accept wastes from out of province.

Other Market Efforts

Other provinces and states, large and small, have encountered similar failures through utilization of some form of market approach to facility siting. In Canada, provinces such as New Brunswick, Newfoundland, Nova Scotia, Prince Edward Island, and Saskatchewan generate small but not insignificant amounts of hazardous waste. Their hazardous waste programs are generally receptive to the idea of private firms pursuing site selection and operation but do little to increase the likelihood that such discussion would begin or succeed. Their basic strategy is continued reliance on either export to various provinces or states or, most commonly, disposal on-site or into municipal sewers and landfills intended for solid wastes.

In the United States, states other than Massachusetts have also generally floundered in their efforts to employ market strategies. North Carolina, for example, has relied upon both market and regulatory approaches simultaneously, with little success. Under the former approach, a private firm can announce its preferred site, apply to the state Solid and Hazardous Waste Management Branch for a waste management permit, and seek local approval of siting.[25] As in Massachusetts, compensation was assumed to be sufficient to entice local cooperation once the site candidate was announced and negotiations ensued.

In practice, however, the North Carolina experience has been a replay of other states and provinces. In 1982, for example, Chemical Waste Management, a private waste management firm, used the market approach in attempting to open a facility in Greensboro. As policy analyst Frances M. Lynn noted, Chemical Waste Management "took an option on land in a residential neighborhood next to a shopping mall. They announced plans to build a facility. For the next 18 months they provided little information to anxious neighbors and other concerned citizens and even forced a TV crew off the site." [26] In response, citizens organized effective opposition and eventually were successful as the state rejected the firm's permit request.

Even firms with public relations skills superior to those demonstrated by Chemical Waste Management in Greensboro encountered similar difficulties with market strategies. Michigan has made several modifications in its hazardous waste facility siting law since the late 1970s but has retained throughout the principle that waste management firms launch siting proposals. [27] In many respects, the impetus for siting is less severe in Michigan than other states and provinces, such as Massachusetts, North Carolina, and British Columbia, because of the existence of treatment and disposal facilities that were opened before the expanding concern over siting and the incremental expansion of some of these earlier facilities. However, controversy has surrounded more recent proposals, including an initiative advanced by Envotech Limited Partnership to open a massive waste disposal facility in Augusta Township, fifteen miles south of Ann Arbor.

The Envotech proposal was introduced in 1987 and called for the construction of a 7 million ton hazardous waste landfill and a rotary kiln incinerator with a capacity of up to 175,000 tons per year. If opened, it would be the second-largest hazardous waste facility in North America, ranking only behind the massive — and continually controversial — WMX Technologies (formerly Waste Management) landfill that was opened in Emelle, Alabama, in 1978 and accepts waste from forty-eight states. [28] The proposed Augusta Township facility is planned for an 1,800-acre site, which includes a 27-acre solid waste landfill that suffers from such massive hazardous waste contamination that it ranks fourth on the state's Department of Natural Resources statewide list of 3,000 contaminated sites. Part of the attraction of the facility was that Envotech would secure cleanup of all contamination, in its plan to develop a comprehensive waste management facility that would serve Michigan and surrounding states and provinces. Economic compensation offers, including job commitments to area residents, were expected to provide further enticement.

Any hope that this "risk substitution" strategy would work in Michigan has dimmed in the face of relentless local political opposition.[29] Groups such as the Augusta Environmental Strategies Committee and Milan Citizens Against Toxic Substances have provided effective opposition, and local governments have raised more than a quarter of a million dollars from bond issues and general revenues in attempting to fight the facility. Envotech has yet to devise a method for fostering public dialogue on the proposal. It has remained embroiled in a series of controversies concerning management of the contaminated landfill and its other area facilities, as well as its overall candor with the public. A major source of concern has been the issue of out-of-state waste exports, as the proposed facility is located about one hour by truck from the Ontario and Ohio borders and is widely assumed to be a magnet for wastes from many areas if opened. Nearly two-thirds of the wastes received from a nearby landfill owned by the firm come from outside Michigan. Envotech announced its abandonment of the incineration portion of its proposed facility in July 1993, although it remains engaged in battle with local opponents over the massive landfill. Once again, the promise of generous, multifaceted compensation packages has proven insufficient to produce siting agreements.

Regulatory Approaches: The Limits of Preemption

State and provincial governments have also experimented in recent years with more aggressive siting methods. Instead of delegating siting decisions to the marketplace, many subnational governments in both Canada and the United States have authority to attempt to impose siting decisions on communities. Under this approach, siting of hazardous waste treatment and disposal facilities is similar to traditional methods used to site highways, housing complexes, and prisons. Legal tools of preemption and eminent domain are often employed, replacing compensation or extensive measures of public participation with strong-arm tactics to impose unpopular facilities on individual communities. Hard-nosed political leadership is thought essential to successful implementation of such a strategy. One leading analyst of the siting process contended that it is essential for the elected executive to insist: "This facility represents a risk that grownups willingly bear for the common good. It isn't abusive, or exploitative, it won't give your children cancer, and we need it. Grow up." [30]

At least ten states and two provinces have relied on such strong-arm, regulatory strategies since the mid-1980s, although some of these operate in addition to market-oriented strategies. In these cases, governmental officials use technical and social criteria to pinpoint preferred sites and then employ a variety of preemptive mechanisms (or, in some instances, enticements). Once sites are selected, state or provincial governments may either choose to construct and operate a treatment and disposal facility on their own or delegate that responsibility to private waste management firms.

Regulatory approaches to siting are guided by what political scientists Bruce A. Williams and Albert R. Matheny have termed managerialism, which operates under "the assumption that organizationally-based expertise can provide the basis for the rationalization of social regulatory policy." [31] Managerial or regulatory approaches thus depoliticize procedures such as siting by substituting expert judgment for public deliberation or decisionmaking by elected officials. They presume that no entity can rival state or provincial agencies for understanding waste management needs and establishing preferred site location. For states and provinces eager to avoid prolonged Nimby clashes and demonstrate an ability to do something concrete about hazardous waste management, these strategies continue to hold considerable appeal. Hence they have continued to be utilized in British Columbia and Ontario and states such as Arizona, Florida, Georgia, New Jersey, and New York.

Regulatory approaches are highly suspect from the standpoint of democratic theory. Williams and Matheny explained that such siting strategies defy traditional pluralistic understandings of regulatory processes, often short-circuiting public debate in search of a siting deal.[32] Market strategies may be somewhat more consistent with democratic theory than regulatory ones, as private firms are at least forced to negotiate with individual communities instead of using preemptive mechanisms to thwart opposition.

Beyond these broader concerns, regulatory approaches are also flawed from the standpoint of their performance. Political scientist William T. Gormley, Jr., noted that selection of "muscular" measures has become increasingly common in regulatory policy, particularly in the United States, and often delivers disappointing results.[33] This pattern is confirmed by the experience of regulatory siting strategies in hazardous waste facility siting, for as policy analyst Jean Peretz noted, "There are no state-sited hazardous waste management facilities in operation" in the United States.[34] A similar observation can be made for Canada. In 1982 policy analysts David L. Morell and Christopher Magorian argued convincingly that it was a myth to assume that preemptive powers could be used to secure siting agreements, given the political and legal tools available to communities so targeted.[35] Policy analyst Christopher Duerksen's examination of industrial facility siting in the same period further indicated that heavy-handed governmental tactics were likely to backfire.[36]

Recent experience has confirmed that states and provinces that attempt to impose sites do so at great political risk. The public reaction is often so strong that specific proposals are abandoned and a fundamental policy shift later ensues. The following sample of cases reveals the extent to which regulatory strategies tend to exacerbate public distrust rather than foster siting agreements.

Ontario

Like a number of American states, Canada's most populous and industrialized province turned to a regulatory siting strategy out of desperation, following repeated failure of market-based approaches. At least five major projects were rejected in the late 1970s, including an incinerator and deep well disposal site, triggering growing concern about the province's capacity to assure sophisticated treatment and disposal. Aside from a privately run incinerator at Sarnia, operated by Tricil, and eight landfills of modest size, many Ontario waste generators remain reliant on such highly suspect disposal methods as municipal sewers and solid waste landfills or exports to Quebec and the United States.

Provincial officials responded in 1980 with a decision to move directly into the business of site selection as well as facility construction and operation. The Ontario Waste Management Corporation (OWMC) was formed, given a charge of securing a site and undertaking construction of a facility that could initially handle 150,000 tons of hazardous waste per year and could double this capacity over a ten-year period. The proposed facility was to feature a variety of waste management technologies, including a rotary kiln incinerator, a physical and chemical treatment plant, and an engineered landfill. Upon reaching its capacity, the facility was expected to handle nearly half of the estimated 800,000 tons of provincially generated waste that must be managed off-site each year, leaving the 3.2 million remaining tons to be handled on-site.

The OWMC began its work with great confidence and quickly selected a site, a 12,500 acre area known as South Cayuga, which had once been promoted for development as a provincial new town. The crown corporation completed its selection process only a few months after its creation and expected to move expeditiously into the facility construction phase. Little provision was made for public participation in the stages before site selection, as is customary under a regulatory siting strategy.

The siting process, however, proved to be far more contentious than the

OWMC had envisioned. Rumors of an impending decision began to circulate in the South Cayuga area and an opposition group, the Haldimand-Norfolk Organization for a Pure Environment (HOPE), formed two weeks before the November 25, 1980, site selection announcement. HOPE was joined within days of the site selection by an array of additional groups, representing churches, area municipalities, and local merchants, among many others.[37] As they probed the site selection process and the suitability of the proposed site for a massive facility, they found numerous problems, all of which reflected poorly on the judgment of the OWMC. The proposed site was located in a floodplain, contained a gas well, and was considered permeable to leakage of hazardous materials into a major groundwater aquifer located below. A hydro-geological study revealed that the South Cayuga site was only "marginally suitable," further indicating that the site had probably been selected on the basis of political factors, including the anticipated absence of opposition, and not on ones of technical suitability.[38]

In response, the OWMC withdrew its proposal in September 1981. This led to a celebration party and a ringing of church bells in South Cayuga but did not augur well for the future of Ontario's fledgling regulatory siting approach. OWMC officials contended that the rejection of the South Cayuga site "constituted a strategic choice to gain environmental credibility for the newly instituted Crown Corporation and to facilitate its future actions." [39] Subsequent efforts to bolster its image involved development of a far more elaborate set of criteria for site consideration and selection. More than 150 separate factors and indicators of site suitability were developed, including hydrogeology, atmospherics, sensitive ecosystems and rare species, archaeology, transportation, agricultural and soil-sensitive crops, tourism, municipal finance and emergency services, number of residents, and family composition.[40] These were to be utilized in a five-phase process that would ultimately culminate in site selection.

The OWMC quickly narrowed its site selection focus to the Golden Horseshoe region in southeastern Ontario. Approximately 70 percent of the province's waste is generated in this area, thereby minimizing waste shipping distances. In addition, the area was located adjacent to the American border, leaving open the possibility of accepting American wastes should Ontario generators prove unwilling or unable to utilize fully any new provincial facility. After phases one and two were completed — essential data gathered, revisions made in the proposed technologies to be used at the site, and a decision reached to concentrate on the Golden Horseshoe area — phase three nar-

rowed the number of candidate sites. Then in phase four came the most important announcement, the selection in 1985 of West Lincoln Township,
Region of Niagara, as the preferred site.[41]

Unlike the South Cayuga case, the proposed siting host had some advance
warning as it had appeared on earlier lists of potential candidates. Moreover,
the OWMC had provided some opportunity for community input into the
site selection process and offered "intervener funding" for citizen groups and
individuals to review various OWMC reports and related documents.[42] Nonetheless, the site selection process remained coercive, and the announcement
has not been well received in West Lincoln Township. Local citizens and officials in the township had long ago joined forces with the Regional Municipality of Niagara, the neighboring City of Niagara Falls, and a consortium
of citizen groups known as the Toxic Waste Research Coalition. They have
continuously attempted to thwart the proposal, relying on the tools of local
zoning approval and the provincial environmental assessment process as well
as direct political protest.[43]

Despite this opposition, the OWMC has remained adamant that the West
Lincoln site is optimal. The corporation rejected an offer from the Ontario
city of Thorold to host a facility, emphasizing the high technical suitability of
West Lincoln. However, the OWMC has already invested more than $100
million in the siting process, for staff support, technical assistance, and intervenor funding, among other expenditures, and has only a protracted Nimby
battle to show for it. Some observers remain confident that the facility will be
opened eventually, although it is likely to proceed only if legally imposed on
communities that clearly do not want to host it.

Regardless of the final siting outcome, the regulatory approach taken in
Ontario has proven extremely costly and divisive. So much time has transpired
since the OWMC began its siting efforts that some have questioned whether
the proposed facility is still suitable for provincial waste needs. The Ontario
experience is widely derided in other provinces as a model of what not to do in
siting, and consequently no other province has adopted such a pure regulatory
approach since Ontario's decision to do so in 1980. As one prominent Manitoba official noted, "When they [crown corporation officials] were finished,
they told [nearby residents] you're the lucky site. Then all hell broke loose." [44]
Manitoba opted to avoid the regulatory approach of Ontario and instead
looked west for ideas, ultimately achieving a siting agreement through its
adoption of the Alberta approach. And, ironically, Ontario appears on the
verge of resolving the sticky question of low-level radioactive waste facility sit-

ing, having devised an Alberta-like voluntary approach instead of the regulatory one it has attempted with such mixed results for hazardous waste.

Florida

The political cultures, economic bases, and ethnic mixes of Ontario and Florida are about as distinct as their respective climates. Yet the two shared a common experience: both abandoned flawed market approaches to hazardous waste facility siting in favor of regulatory ones and faced enormous public opposition when final siting decisions were announced by a governmental authority.

Florida generates about the same amount of hazardous waste, with about the same portion going to off-site facilities, as Ontario. However, it has minimal commercial treatment or disposal facilities and has become heavily reliant on export, particularly to facilities in Alabama and South Carolina, for off-site disposal. State government and industry officials have long acknowledged the need to develop some treatment and disposal capacity and decided to turn to a regulatory approach after repeated failures of market-style siting efforts. The growing chorus of complaints from the hosting states and the possibility of EPA sanctions if Superfund capacity assurance commitments could not be met further prompted this new course.

In 1987 the state amended its waste facility siting legislation to require the Florida Department of Environmental Regulation to recommend a site for a multipurpose hazardous waste treatment facility to the governor and legislature by May 1, 1988.[45] This approach also was expected to lead to the development of smaller storage facilities around the state. The legislature felt that state agency action was necessary to ensure a successful choice of sites and that a deadline was required to force action before the requirements of any capacity assurance report had to be met. Successful site selection presumably would ease the process of securing the services of private firms to construct and manage the facility; unlike Ontario, this regulatory form of site selection was intended to lead to private, not public, facility management.

The site selection process began in 1987, as the state required each county to recommend areas in which a "storage facility could be constructed to meet a demonstrated need." In addition, each regional planning council was required to "designate one or more sites at which a regional hazardous waste storage or treatment facility could be constructed." [46] These steps were intended to build a sense that the burden for waste management would be shared broadly across counties as well as to build public awareness of the need for storage, treatment, and disposal facilities.

The process proved far less contentious than the evaluation, by the Department of Environmental Regulation (and its consulting firm, Roy F. Weston, Inc.), of potential sites for a comprehensive hazardous waste facility for all of Florida. For this facility, the state limited site consideration to only those areas that were either owned by the state, privately held but donated to the state for the express purpose of hosting a comprehensive facility, or volunteered to the state by industrial parks or other land holders receptive to hosting a facility.

Narrowing the pool of potential sites raised some questions, particularly given the relative paucity of promising sites because of pervasive groundwater and related hydrogeological concerns. Nonetheless, state officials proceeded with an evaluative process that in many respects paralleled the one developed by the Ontario Waste Management Corporation. Efforts were made to inform public officials of possible siting areas in or near their boundaries, but the review process attracted little attention until the state had selected five site finalists in January 1988. Two of these were in Union County, and one each was in Bradford, Clay, and DeSoto Counties.

Speculation immediately arose that political and economic factors, not technical considerations, had driven the site selection process. Each of the final sites suffered from serious economic problems, and several of them had substantial racial minority populations, thus confirming a traditional pattern to concentrate siting in the areas presumably least equipped politically to respond. The final site selection, announced May 1, 1988, targeted state-owned prison land in Union County, which is southwest of Jacksonville. The county is one of the smallest and poorest in Florida, with the smallest tax base and a budget deficit beset by a series of illegal deficits for many years.

The presumed vulnerability of Union County notwithstanding, a strong Nimby reaction emerged and ultimately resulted in the state's withdrawal of the site proposal. Within a week of the final recommendation, a new local advocacy group called United Citizens Against Pollution (UCAP) filed a petition for a formal administrative hearing, claiming that the process used by the state was "faulty and politically biased." Unlike Ontario, Union County was not entitled to intervenor funding and had no fiscal reserves to mount a defense. Nonetheless, the group raised $30,000 — through fish and chicken dinners, bake sales, and individual contributions — to hire a lawyer and environmental consultant.

UCAP received little support from nationally based environmental organizations and little attention in the most prominent state media outlets. However, its opposition was so strong and convincing that in 1991 the Department of Environmental Regulation decided to shelve the Union County proposal. The site, technically, is "being held in reserve" in the event that other facilities being proposed under the continuing regulatory strategy option fail to materialize.[47] Some contend this is merely a delaying tactic, assuming the state hopes that continuing economic decline will ultimately convince Union County opponents to change their views. However, much as in Ontario, this top-down regulatory approach to siting appears doomed to failure.

In fighting the proposal, UCAP discovered that the Weston contracting firm had not carefully implemented the state's plan for site selection. It lacked essential documentation, such as basic maps, despite its claims to the contrary. As an environmental lobbyist noted, the Weston firm had "spent a lot of money and screwed things up royally," thereby confirming the suspicion of many in Union County that they were being targeted out of perceived weakness instead of out of application of any fair, objective siting criteria.

New York

The Florida experience with regulatory approaches to siting has been matched by other states adopting a similar approach. Even when states not only attempt to select a site but also hold out the possibility of constructing and operating a facility on their own, the Nimby response has become commonplace. New York proves, in this regard, to be no different than neighboring Ontario.

In New York, prior failures of a market strategy compelled Governor Hugh L. Carey, the state legislature, and the state Department of Environmental Conservation to embrace a plan for state selection, construction, and management of a high-capacity hazardous waste incinerator that would be located on state-owned land. The state intended to use the Environmental Facilities Corporation, a public authority created in 1970 to assist with the construction and management of pollution control facilities, to oversee the site selection and development process. This authority and the general approach used in New York was consistent with a long-standing practice in the state "in the use of state-chartered authorities to gather private investment for a wide variety of politically controversial projects." [48]

Construction of a major incinerator, however, was to prove far more difficult than the grand construction efforts of earlier state power brokers such as Robert Moses and Nelson A. Rockefeller. A technically driven site selection process led officials to declare a site in Sterling owned by Rochester Gas and

Electric as their preferred site. The state wanted to purchase 400 acres of the 2,800-acre area for the facility, which was located eight miles southwest of Oswego and borders Lake Ontario. Earlier reports indicating Sterling was a potential candidate triggered some local opposition, and these early signs of conflict exploded shortly after the state announcement. The towns of Sterling and neighboring Fair Haven adopted resolutions condemning the proposed project the same day as the announcement. They were supported in following days by formal expressions of opposition from the Oswego County legislature and the Central New York Regional Planning and Development Board.[49] In turn, a grassroots environmental group, which had formed in 1973 to oppose a nuclear power plant that had been proposed for the area, reemerged and offered Nimby-type resistance.

All of this opposition was crucial to a November 1981 decision by the state to drop the Sterling site from further consideration, although it also acknowledged difficulties in securing the land from the power company and uncertainty in demand for the facility from waste generators.[50] Since this siting setback, the state has abandoned further regulatory efforts and has emphasized market strategies as well as regional agreements to secure access to treatment and disposal facilities. These efforts have been largely unsuccessful, although New York has been somewhat effective in reducing the need for added facilities through aggressive waste recycling and reduction efforts.[51]

North Carolina

By the mid-1980s it became obvious that the market approach to siting taken by North Carolina was flawed. No new siting agreements were in view, and states responsible for taking substantial waste exports from the Tar Heel state were complaining loudly to Washington, D. C., that some serious steps toward siting by North Carolina were in order. In response, the General Assembly of North Carolina created a Hazardous Waste Facility Siting Commission in 1984. This represented an endorsement of a regulatory strategy, putting the state directly into the process of site selection and, possibly, facility operation, although it kept open the option of a market-based approach as well.

The commission consisted of nine members, three each appointed by the governor, the assembly upon recommendation of the Speaker of the state house, and the assembly upon recommendation of the lieutenant governor. Given the inherent complexities of siting, the marching orders given the commission were bold. Appointed in early 1985, the commission members were to develop "a comprehensive plan for the treatment of hazardous waste in North Carolina, including a plan to provide for a state hazardous waste collection system" by May 1, 1985. In the event no permit had been issued for operation of a private waste facility operator by June 1, 1985, the commission was to "actively seek communities interested in hosting hazardous waste treatment facilities and private operators of hazardous waste treatment facilities and shall present appropriate sites . . . to these operators." [52]

After this, if no permit for operation of a facility had been issued to a private operator by January 1, 1986, the commission was empowered to "select appropriate site(s) and begin proceedings to purchase or if necessary condemn property for such site(s) under the state's power of eminent domain." Finally, in the event no permit to operate a facility had been approved by June 1, 1986, the commission was to submit plans to the General Assembly "for construction of a facility on one of the sites and shall proceed to begin construction of a facility within one year and shall seek a private operator to operate the facility." After this, if no private operator could be secured, it would be up to the commission to operate the facility. [53]

Such a legislative charge gave state officials enormous latitude and authority in attempting to resolve North Carolina's hazardous waste treatment and disposal dilemma. However, the broad delegation of authority did not translate into a workable siting program. The commission attempted to adhere to the tight timetables and did little to educate and engage the public in prolonged dialogue over the need for a facility and siting alternatives. Instead, the commission devoted enormous attention to trying to find an optimal site, reviewing more than 500 options before settling on a 2,400-acre site in poor, rural Lee County in the spring of 1988. The commission proposed construction of a comprehensive waste management facility, including a pair of rotary kiln incinerators with a combined capacity of 35,000 tons of hazardous waste per year. A strong Nimby reaction ensued from Lee County residents, prompting the General Assembly to veto the selection. The assembly also told the commission to pursue sites only in counties that had volunteered, contrary to its early advice to pursue siting expeditiously and with a minimum of public input. [54] In 1989 legislative reforms, the assembly also renamed the organization as the Hazardous Waste Management Commission and directed it to halt further siting efforts until a thorough study could be completed on North Carolina's hazardous waste management needs. [55] However, much of the early activity of this newly formed commission involved refinement of technical criteria for regulatory-style site selection, which by the early 1990s was beginning to focus on two areas, the Iredell-Rowan county line, located between the cities of Salisbury and Stateville, and a tract in Granville County, located near the town of Oxford. Observers concur that distrust runs high and that any siting agreement is a long way off. This continuing standoff and the enduring reliance on hazardous waste exports have made North Carolina the first state to face the possibility of formal cutoff of federal hazardous waste cleanup funds.[56] To date, however, EPA has backed away from taking such steps.

New Jersey

No discussion of hazardous waste in the United States and Canada would be complete without a review of the state that has become most synonymous with the issue of hazardous waste. Much of the existing American federal regulatory infrastructure, including Superfund and community right-to-know legislation, drew heavily from New Jersey's legislative actions in the 1970s and 1980s.[57] New Jersey's problems related to hazardous waste were so widespread and dramatic that they triggered considerable policy innovation, ranging from mandatory environmental audits on transaction of commercial property to mandatory toxic emergency evacuation plans that must be filed by major corporations.[58] Many of these problems remain; for example, New Jersey continues to rank atop federal listings (such as the National Priorities List of the Superfund program) of the states with the largest number of serious abandoned hazardous waste facilities. The state also remains one of the nation's largest generators of hazardous waste.

Unlike Massachusetts and North Carolina, New Jersey does have a fair amount of treatment and disposal capacity off-site. However, state officials acknowledge that added capacity is needed, especially given the uncertainty of export markets. As of the mid-1990s the state was exporting more than 200,000 tons of hazardous waste per year to other states and also was exporting wastes to a large variety of developed countries (such as Canada) and undeveloped ones (in Africa and Latin America). Consequently, state policymakers devoted considerable attention to the issue of facility siting and wavered between traditional market and regulatory strategies, much like North Carolina. Their considerable efforts in this area have generally failed, fanning an intense Nimby reaction in case after case.

Part of New Jersey's multifaceted attempt to address its hazardous waste management problems involved enactment of the 1981 Major Hazardous Waste Facilities Siting Act. This act provided for the planning, siting, and licensing of new commercial waste facilities and established activities to be carried out by the Department of Environmental Protection, the Hazardous Waste Facilities Siting Commission, and the Hazardous Waste Advisory

Council. This process began in 1981 with development of siting criteria and related regulations. The state received help from environmental groups such as the Sierra Club in the drafting of these materials and hired as commission executive director Richard J. Gimello, a former citizen activist who played a major role in drafting the siting act.[59]

Although the state was successful in incorporating large environmental groups into the early planning process, it was less committed to building trust at the local level. The commission chose to select sites on technical merit instead of social or political considerations or any voluntary process. Eleven sites were announced on February 14, 1986, which many would later remember as the "St. Valentine's Day massacre." As journalist Carl LaVo reported, "Within hours, general support for the state effort in the 11 towns was transformed from fear to rage."[60] Many local groups sprang up in opposition that ultimately thwarted development of any of the sites. The state had undertaken a major public information campaign before the announcements, but these made a general case for a state-based facility and had not prepared individual communities for the possibility of serving as host. And as is often the case under market strategies, state efforts to dangle generous compensation packages, including 5 percent of the gross receipts from a facility, were insufficient to win support in any community.

In Burlington Township, for example, one of the eleven finalists for siting, 4,600 citizens attended a public meeting at a local middle school, 3,000 of them forced to watch on closed circuit television because of gymnasium overcrowding and 1,300 more turned away at the door. The vast majority of attendants jeered Gimello and other siting commission members as they attempted to make a case for siting in Burlington. Typical of such proceedings, in New Jersey and elsewhere, the stage backdrop consisted of a skull and crossbones that marked an aerial photo of the proposed site, with bull's-eye rings indicating the distance of schools, water supplies, and other communities to the targeted site. The hallway near the Burlington Township school gymnasium was packed with students' artwork that depicted their fears over the possibility of an incinerator being located in their community.[61] Given such opposition, Burlington Township was quickly withdrawn from further siting consideration, as were the other ten sites initially selected by the siting commission. The siting process has remained gridlocked in New Jersey.

Other Regulatory Efforts

Variants on the New York and Ontario approach that take a more extensive regulatory role than site selection have fared equally poorly. In Georgia, an initial effort to pursue siting of a state-sponsored facility in a poor, rural county was blocked when charges of political influence in the site selection process were proven accurate. A subsequent effort to impose a site upon rural Taylor County, thirty miles east of Columbus, has faced considerable local opposition and is at a point of impasse. In Arizona, state officials invested more than $50 million in trying to site a massive state-sponsored hazardous waste incinerator in Mobile, thirty miles southwest of Phoenix. Local opposition was exceptionally vehement, so much so that Maricopa County sheriff deputies used stun guns to subdue protesters at a local meeting.

The Common Nimby Outcome

None of the cases presented in this chapter is identical and yet they fall into two decided patterns that arrive at a single common outcome: siting gridlock and proposal rejection. Provinces and states that have relied on either market or regulatory approaches have failed to foster the trust necessary for prolonged deliberation over the possibility of site acceptance, much less moved to the stage of final agreement. For all their differences, the purported strengths of each strategy succumb to similar, Nimby-type reactions. In the case of market approaches, compensation never becomes much of an enticement. It is never seriously explored in many cases, because local animosity toward siting proponents is so great. In the case of regulatory approaches, governmental authorities are rejected as having employed arbitrary criteria to deposit the waste disposal burden of a state, province, or region on a single community. In many cases, the target communities are poor and populated with large numbers of racial minorities, thereby fueling the sense of inequity in distributing these burdens.

For all their differences, these cases are linked not only by the resulting political conflict and proposal rejection but also by the failure of private (in market cases) or public (in regulatory cases) officials to develop mechanisms through which genuine public discussion and exploration of waste management options can occur. Siting agreements become feasible only after an atmosphere of trust has been established, when individual communities no longer feel dumped upon for arbitrary reasons but instead are part of a larger, multifaceted approach to waste management that involves multiple communities and broad sharing of burdens. This process is not an easy one, especially given the profound distrust that surrounds siting after a generation of failed efforts. But alternatives do exist to traditional siting approaches. And in at least a few instances, particularly but not exclusively in Canada, a very different approach to hazardous waste facility siting and waste management can be examined, one that operates through a process of collective deliberation and pursues facility siting only in communities that have been well appraised of benefits and dangers and subsequently agree to serve as site hosts. Such processes are not perfect, but they provide some tangible evidence that classic Nimbyism need not be the only outcome of hazardous waste facility siting in Canada and the United States.

3

The Alberta Case

THE STEADY DRUMBEAT of siting failures of hazardous, radioactive, and biomedical wastes calls into question the very possibility of facility siting agreements being reached. Given the shortcomings of prevailing market and regulatory approaches, siting gridlock may appear an inevitability. Although some salutary effects of the Nimby phenomenon should be acknowledged, the virtual inability to open new hazardous waste treatment and disposal facilities in either Canada or the United States raises a host of serious public health, environmental, equity, and intergovernmental concerns.

A growing theme in the rapidly expanding literature on hazardous waste facility siting is the expectation of a Nimby outcome, whether condemned or celebrated. On occasion, proposals are put forward to try to reverse the Nimby trend, but these are often made at a general level of abstraction and are based on little or no actual experience, either in North America or elsewhere. Meanwhile, specific, exceptional cases do exist; siting agreements have been reached. In some of these cases, new sites have been approved and new facilities put into operation, with a wide base of community support. Though relatively few in number, they may point the way to a more effective approach to facility siting—and hazardous waste management—in the years ahead.

The exceptional cases that lead to siting agreements defy any monolithic pattern; they offer no easy blueprint for a state or province to follow that guarantees siting success. Each case, however, features several shared design characteristics integral to each agreement. These characteristics played an important role in gaining the public support necessary for successful siting. When integrated, these features render a siting process very different from those developed in the states and provinces using market approaches, regulatory approaches, or a combination of both.

The successful siting cases are linked by two special types of assurances given to communities against unfair treatment by private firms or governmental agencies with responsibility for waste management. First, siting is advanced as part of a voluntary process; that is, it is considered only in technically suitable settings among communities that volunteer to host treatment or disposal facilities. No coercion takes place that is so fundamental to market or regulatory strategies, whereby targeted host communities have no prior knowledge of their select status. Under the voluntary approach, siting agreements can be reached only after extensive public information sessions have been held and ample opportunities have been provided for citizens to raise questions and determine their support for or opposition to the proposal. The deliberations must cover the institutional arrangements that will be made to assure safe facility construction and management, compensation packages to the host community, and long-term measures providing community oversight of facility operation. The intent is to build trust, extend discussion, and foster a sense of genuine deliberation over the possibility of siting instead of submitting a take-it-or-leave-it siting proposition that most likely would be rejected.

Second, further reassurances are offered against exploitative practices. Perhaps the most unfair aspect of facility siting, alongside the often arbitrary nature of market or regulatory site selection, is the concentration of a massive waste management burden upon a single state, province, or community. All too often, the siting of treatment and disposal facilities has operated independently of other aspects of an effective waste management system. The concept of burden sharing that would systematically link one or more facility siting efforts with construction of regional transfer sites and expanded emphasis on waste recycling and reduction in all industries and communities is extremely rare. In turn, communities in Canada and the United States rightly fear that, if they agree to host a major treatment or disposal facility, broad waste management burdens will be deposited squarely on them.

Should a community offer unilateral cooperation, under traditional siting arrangements, other communities will be freer to reject waste facilities. In addition, regional waste storage and transfer stations, household hazardous waste programs, and massive recycling and reduction efforts will become less desirable. Prospective site hosts fear that they will draw wastes not only from nearby industries and neighborhoods but also from other states, provinces, and nations. In effect, under prevailing siting strategies, communities face considerable incentives to refrain from cooperative collective action. For if they do not, they will have to bear any environmental or economic risks that stem from hosting a waste management facility, and they will have to assume a burden that might be transferred elsewhere or otherwise be shared more equitably.

More successful siting programs emphasize voluntarism as well as a commitment to avoid concentrating waste management responsibilities on a single site host. Facility siting thus becomes only one part of a multifaceted approach to hazardous waste management that places strong emphasis on broad governmental, industrial, and citizen commitment to a waste management partnership. Every citizen does not have to become an environmental activist and radically change lifestyle and consumption patterns. And the individuals and institutions responsible for generating the greatest volumes of waste should have to bear the greatest financial burdens in assuring safe waste management. However, a broadening of the responsibility for hazardous waste management needs to occur, recognizing the ubiquity of sources that generate hazardous waste in modern societies. Hazardous waste is thus treated as a collective good for which collective responsibility must be taken. Furthermore, formal assurances can be given to a community hosting a treatment or disposal facility that it will not become a magnet for wastes from other states, provinces, or nations that have been unwilling or unable to create a partnership of their own.

American and Canadian governments and waste management firms are generally held in low esteem by the public. And more than a decade of failed market and regulatory approaches to siting, not linked to larger assurances or waste management strategies, has only exacerbated these tensions. However, a series of recent cases suggests that an alternative approach to siting can be devised that, at least in some instances, can be successfully implemented. From Alberta's experience and several cases in both nations involving hazardous waste and low-level radioactive wastes emerge shared design characteristics that facilitate replacement of the typical Nimby response with a siting agreement, substantial movement toward successful completion of an accord, or creative waste reduction techniques.

Beyond the Nimby Syndrome in Alberta

The western province of Alberta opened the Swan Hills Special Waste Treatment Centre in September 1987, following a radical transformation of the provincial siting process. Site proposals advanced by private corporations in the late 1970s met with outraged local resistance in the communities of Fort Saskatchewan and Two Hills, and the prospects for any future agreement seemed remote. In response, the Alberta Ministry of Environment declared a moratorium on future siting and established a committee to review the provincial siting process and propose reforms.

A new process resulted, which involved the province in novel approaches to siting. Alberta established a crown corporation (a publicly held enterprise, common in Canada) to manage any major waste facility jointly with a private corporation that would be selected by the province. It also developed a comprehensive program for educating the public and considered siting only in communities that met provincial environmental standards and volunteered as site candidates. After a multilayered process of consultation with varying local governmental jurisdictions, five communities came forward as potential site hosts. Swan Hills, a community of 2,396 residents located 209 kilometers northwest of Edmonton, was selected in 1984, after a plebiscite in which an overwhelming majority of its voters supported the facility proposal.

The Swan Hills Special Waste Treatment Centre is the most comprehensive waste treatment and disposal facility in North America. It can incinerate organic liquids and solids, treat inorganic liquids and solids, and provide a landfill for contaminated bulk solids. It has a potential capacity of more than 100,000 metric tons of hazardous waste annually, adequate to cover the vast majority of hazardous wastes generated in Alberta that require off-site management. The center has undergone some expansion of its incineration capacity in 1993-94, as a result of the unexpected discovery of large quantities of organic wastes that had been dumped inappropriately around the province.

The center is designed as a central part of a comprehensive hazardous waste management program for the province, which includes regional transfer stations, an extensive program that promotes waste recycling and on-site waste reduction, and unusually strict controls on wastes transferred by vehicle. The provincial effort to reduce the total volume of waste requiring treatment at the center is second only to Manitoba as the most far-reaching of any Canadian province. It fully rivals the most innovative American states in this regard, making prevention a central part of the provincial waste management strategy.

The center continues to enjoy widespread support among local residents and provide extensive opportunities for public oversight of the facility and access to all of its functions.

Alberta and Its Waste Management Problem

Like other Western provinces, Alberta has hardly been known for its abiding commitment to environmental protection. Unlike many American states and Canadian provinces such as Ontario, Alberta has devised a series of social regulatory policies intended to minimize governmental interference with economic development. Because so much of the province's post–World War II growth has been derived from the exploitation of its natural resources, particularly oil, gas, and timber, authorities have been reluctant to interfere with these industries.[1]

Such an economy has led to a pattern of boom-and-bust development for the province, one that rides particularly high as demand for its raw materials intensifies. Provincial governments have wavered from active support of industrial expansion, including investments in training, highways, and tax subsidies, in the 1970s, to more cautious fiscal strategies in the 1980s and 1990s.[2] Unlike other provinces, however, Alberta has the considerable cushion of the Alberta Heritage Savings Trust Fund, a multibillion-dollar fund available for public investments drawn from nonrenewable resource revenues of the late 1970s and 1980s.[3]

These reserves were insufficient to guard against economic decline during the early 1980s. Alberta's unemployment rate reached double digits, remaining as high as 10.6 percent as late as 1986, when recovery was well advanced in other regions. Nonetheless, it was during this period that Alberta decided to begin to address an increasingly serious problem: hazardous waste.

By the early 1980s Alberta ranked fourth among Canadian provinces in hazardous waste generation, producing an estimated 215,944 tons that placed it narrowly behind British Columbia. Like other estimates of the period, this one probably was an underestimation. A 1991 study projected that the province would generate 325,510 tons in 1992, even after several years of significant waste reduction efforts.[4] Alberta has a population of nearly 3 million, and the majority of its wastes emanated from major metropolitan areas such as Edmonton and Calgary. But the wide dispersal of the oil and gas industry, and other small industrial and service firms that generate hazardous waste, assured a fairly broad distribution. Moreover, despite the vast territorial holdings of the province (more than 255,000 acres), a major, largely northern, section was ruled out for facility siting because of permafrost conditions that make landfilling virtually impossible and transportation treacherous.

Before the opening of the Swan Hills facility in 1987, Alberta lacked any advanced system for hazardous waste storage, treatment, or disposal. Officials had little idea of how much waste was being generated (making estimates from the period precarious), much less where it was being disposed.[5] One of the initial provincial studies on the issue, published in 1979, concluded that "until now the problems of disposing of industrial waste, except for some minimal provincial regulations and municipal by-laws, have been left almost exclusively to the generators of the waste. It is routinely concealed as normal waste and disposed of in unsuitable areas."[6] This practice was neither unique to Alberta nor the preferred method of disposal. In urban areas, sewers and solid waste landfills were frequent dumping grounds. In one common practice, trucks carried hazardous waste into coin-operated car washes and then flushed their cargo into holding pits that would eventually enter sewers or landfills.[7] Even outlying areas, where considerable gas and oil extraction occurred, were not insulated from these problems. A newspaper editor in such an area said, "It's all oil patch here. Absolutely nothing else. These guys work in the industry, and they know where the stuff is going now. It's going into the goddamn ditch sometimes."[8] As a provincial official confirmed, "We have wastes left from the oil and gas industries left all over the province. They didn't give a damn about wastes. Often, they just shoved it into the bush." Some waste management relief was afforded by exports, particularly the shipment of PCB wastes to facilities in Oregon, but waste generators faced few provincial pressures to take even these steps.

Like in so many other states and provinces, Alberta's policy toward hazardous waste facility siting embodied a market approach. Officials were receptive to the idea that Alberta would operate advanced treatment and disposal facilities but perceived the siting process as a straightforward transaction between the waste management firm and the community selected for siting. Two proposals came forward in late 1979, and both were rejected in short order. In Fort Saskatchewan, a town of 12,000 citizens twenty-five kilometers northwest of Edmonton, Kinetic Contaminants Canada proposed construction of a hazardous waste incinerator. After the site was announced, an initial public meeting was scheduled for mid-September, and it drew a large, hostile crowd. After one resident gained a near-universal show of hands in opposition to the proposal, he said to participating Kinetic officials, "Sir, I think we have summed up the situation for you. You are not welcome here. Go and build your plant somewhere else."[9]

The company withdrew its proposal but made a similar offer to the small (population 1,141) town of Two Hills, 110 kilometers northeast of Edmonton, in December 1979. Not unusual for market strategies, Kinetic Contaminants held private meetings with key local officials before any public siting announcement. This secured an endorsement from the Two Hills Chamber of Commerce and tacit support from the town council before going public. Once the official announcement was made, the public reaction in Two Hills was similar to that in Fort Saskatchewan a few months earlier. It was, in some respects, even more antagonistic once citizens learned that they had not been consulted after extensive meetings between company leaders and local public and commercial officials. Townspeople even threatened to erect tall fences to separate themselves from those neighbors who might support the proposal.[10] Kinetic subsequently withdrew its proposal.

For all its purported concern for natural resource exploitation and economic development, Alberta was not proving to be an easy mark for waste facility siting. By 1980 the province chose to abandon its market strategy and instead conduct an extensive analysis of its hazardous waste problem and management options. It placed a moratorium on any further attempt to site hazardous waste facilities and established a Hazardous Waste Management Committee to study the siting problems and devise an alternative siting and waste management process. A radically different siting approach developed, which emphasized a voluntary strategy.

Public Participation

Creation of meaningful mechanisms for public involvement in environmental decisionmaking proved to be a central component in the Alberta approach. These mechanisms took multiple forms, ranging from highly formal to informal, but all reflected a commitment to securing maximal public participation and voluntary community siting approval before siting decisions were completed. Extended public dialogue resulted, as well as an unusual degree of openness and trust from which serious exploration of options could occur.

Participatory democracy appears more attractive in Canada and the United States as confidence in representative institutions has plummeted. Use of direct democracy tools, such as referendum, initiative, and recall, continues to climb at the state and local level in the United States, especially on environmental and fiscal matters.[11] Similarly in Canada, direct democracy is receiving an increasingly serious examination, having served as the vehicle for rejection of the so-called Charlottetown Constitutional Accord in 1992 as well as for a growing number of issues at the provincial and, particularly, municipal level.[12] But the use of electoral measures can be complex, particularly on issues such as siting in which much of the responsibility for a collectively generated problem must be concentrated on a single community.

Alberta was able to utilize the plebiscitory process with success, but only when combined with a larger set of reforms designed to maximize public participation throughout every phase of the provincial siting process. The venues found for public participation were meaningful and capable of building confidence. By contrast, most states and provinces continue to view the entire issue of public participation in environmental policy with trepidation. They face political, and often legal, requirements to be open in their dealings, but fear that any extensive public role will force them to alter preferred plans or scuttle any prospects for cooperation. As a result, public participation outlets in environmental policy often bombard the public with documentation, much of it technical and incomprehensible, while minimizing any genuine public deliberation or dialogue.

Both public and private sector policy professionals in the area of hazardous waste management and facility siting policy in Canada and the United States have generally approached the question of public participation in various stages of the siting process with pause. Under the market and regulatory approaches they have delayed public involvement until the last possible stages of the process, once a specific site has been announced by either a private corporation or a governmental agency. This proves too late to attain any meaningful public input or build trust between site proponent and the proposed host, much less mount "the systematic and sustained effort necessary to create true democratic dialogue among citizens" in environmental policy.[13]

Alberta, through its Hazardous Waste Management Committee of the early 1980s, went further in establishing an opportunity for dialogue than any other state or province had previously. The committee operated in the absence of any structured process for siting or provincial regulation of hazardous waste management, having been encouraged by the province to design a novel approach. The six-member committee consisted of three provincial officials, including two prominent figures from the provincial environment ministry; a farmer; the chief of the Calgary fire department; and a University of Alberta chemistry professor with expertise in hazardous materials.

In its 1980 report, the committee provided the basic structure of the approach that was ultimately embraced by the Alberta legislature. After review by the Environmental Council of Alberta, an independent review commission, the provincial environmental minister created the Hazardous Waste Task Force to implement the site selection process. This new approach emphasized volunteerism; only communities offering to host a site would be considered as candidates. In addition, private developers would be asked to propose facility plans to provincial authorities, thereby maintaining a market option. At the same time, however, the new Alberta approach established a major provincial role through development of siting criteria and education of the public as to the nature of the hazardous waste problem and alternative remedies. It also was designed to allow provincial authorities to make the final decision on site selection, in the event of community acceptance, and to choose the private corporations that would be involved in construction and operation of the site. The province would also play a direct role in the management of the facility. This blending of features resulted in a systematic role for government in the hazardous waste siting process that was unprecedented.

Siting criteria were applied through constraint mapping, which ruled out parcels of Albertan territory that were deemed inappropriate for various physical, biological, economic, social, and political reasons. Unlike the siting efforts in other provinces and states that utilized constraint mapping, the efforts in Alberta were shaped through exhaustive consultation with the public.

The Alberta approach began with advertising, a potpourri of general informational meetings, and frequent sharing of technical and related reports with community organizations. Communities that expressed an interest in the possibility of hosting a site were offered a detailed — and free — provincial assessment of their area, which could prove useful to them in considering the viability of a hazardous waste site as well as potential landfill sites or other uses. Fifty-two of a possible seventy jurisdictions requested these assessments, and they were invited to volunteer to further explore the possibility of hosting a site.

The province established a number of liaison and other committees that were intended to foster regular and direct interaction between public, provincial, private corporation, and crown corporation representatives at every stage of the siting process. One committee, Alberta Environment, hosted more than 120 meetings, holding at least one in every county, municipal district, improvement district, and special area in the province. Alberta Environment officials responded to citizen questions, provided briefings on the hazardous waste situation in the province, and offered general information on the types of criteria that can be used in a siting program.[14] Those communities that expressed interest in possible participation continued to have far-reaching access to provincial officials and hazardous waste data. According to one Edmonton newspaper columnist, no great fan of traditional provincial approaches to environmental policy, such provisions for public input into provincial decisionmaking were "almost revolutionary in Alberta." [15]

These processes did not, however, assure a conflict-free implementation process. Some of the workshops run by the Alberta Rural Education and Development Association, hired as a consultant by the Hazardous Waste Task Force, were ineffective. Basic lines of communication and jurisdiction between participating governmental units were blurred, and incomplete information was released to the public on occasion. The task force also began to rush the siting process, perhaps because of its self-imposed March 1982 deadline and excessive confidence over an expectedly favorable plebiscitory vote. As a result, Beaver County, which is located 100 kilometers southeast of Edmonton and would later actively seek the comprehensive facility, rejected the proposal in an April 1982 plebiscite by a 1,359-to-556 margin. Early support for the facility was evident. For example, Environment Minister Jack Cookson was presented a petition bearing the signatures of 620 local residents a few weeks before the plebiscite. However, a strong opposition group known as the Bruce-Viking Agricultural Protection Association formed and was effective in challenging the competence and candor of the task force before the vote.[16]

Despite some discussion of abandoning the voluntary approach, task force leaders persisted and were supported by senior provincial officials. The Alberta Rural Education and Development Association was relieved of further duties, and the task force relied instead on a small group of officials with special skill in risk communication to conduct further meetings. "I wore out one and a half cars going from town to town," recalled one official. "I learned what their concerns were." The task force reduced its earlier emphasis on large town meetings, such as the one that went badly in Beaver County, in favor of greater numbers of smaller meetings in which a larger percentage of the populace could participate and vocal opponents would be less likely to dominate. "At first, I was a strident opponent of the plant," explained one local merchant. "And I was the last person who would ever be able to speak up at a big, public meeting. But in the smaller meetings I felt comfortable. I got to ask all of my questions and had them answered. It made all the difference."

These changes revitalized the voluntary siting process and resulted in five communities remaining eager to pursue the possibility of further involvement. All of them held plebiscites in late 1982 that drew heavy voter turnout. All five plebiscites were approved, and three of these were passed by large margins: Swan Hills, Ryley (a town of 500 citizens located in Beaver County), and Veteran (a town of 300 residents located 200 kilometers northeast of Calgary). Seventy-nine percent of Swan Hills voters supported the facility proposal in a plebiscite in which 69 percent of eligible voters participated. Alberta Environment selected the town as the site for a comprehensive waste facility in March 1984. Community leaders from Ryley were extremely outspoken in registering their disappointment in not being chosen. Turning the concept of Nimby on its head, some Ryley citizens responded to the Swan Hills selection by criticizing the government's decision in a full-page *Edmonton Journal* advertisement and by storming the Alberta legislature in a futile protest.[17]

Swan Hills proved attractive to provincial policymakers because it was relatively close (209 kilometers) to the major metropolitan area of Edmonton and directly linked to it by a provincial primary highway, the so-called Grizzly Trail. At the same time, unlike Ryley and some of the other candidate sites, Swan Hills had no immediate neighboring communities, so its acceptance of a facility did not require the support of any nearby towns. Moreover, undeveloped land for a facility was readily available. Speculation endures that local political clout was crucial in tipping the decision to Swan Hills. Ken Kowalski represented the Swan Hills area in the provincial legislature and served in the Progressive Conservative cabinet of Alberta premier Donald Getty. He remains widely heralded in Swan Hills for his efforts to secure the waste treatment center, was prominently featured during the gala opening of the facility in 1987, and has since had a major street in the town named in his honor.

The politics of site selection notwithstanding, the Swan Hills case remains remarkable for its reversal of the familiar Nimby pattern. The extensive public participation process made possible an outcome entirely different from traditional siting cases, as well as the three previous ones in Alberta. Local political leadership played a pivotal role in using the public participation mechanisms to build public trust in the provincial siting process and support for pursuing the waste management facility. Unlike Two Hills, where government officials undermined possible support by behind-closed-doors deliberations with site proponents, Swan Hills officials attempted to move everything out into the open. Leaders emphasized that the siting process was voluntary and that the proposed facility was part of a comprehensive waste management strategy.

Upon initial discussion, many Swan Hills residents expressed alarm and formed citizen opposition groups. "When I brought the idea back to council, I was almost run out of town on a rail," recalled a former Swan Hills elected official. But after the council embraced the idea, a citizens' committee was formed to hold regular public meetings before the plebiscite. These gatherings were held every week over a twelve-week period, and Swan Hills residents were actively encouraged to attend at least two of them. All relevant provincial and local officials were available at the meetings to discuss any aspect of the proposal. "We became taxi drivers, dishwashers, babysitters, whatever it took to get everyone out," said the former official. "We divided up the phone book and called everyone in town."

Such extensive deliberations also served as a forum to consider — and refute — claims from national and international environmental groups that the facility would pose dire environmental and public health consequences if it were accepted. Numerous observers recalled that the efforts of international groups such as Greenpeace to stimulate opposition to the facility proposal were failures. "People from Greenpeace walked up and down the streets of the town looking for money and someone to oppose the facility. They couldn't find much support, so they left," noted one former town official in a typical comment. Even the director of the local environmental group Friends of the Environment in Swan Hills (FRESH), which wins widespread respect for monitoring developments at the facility and promoting safe environmental practices in Swan Hills, was alienated by Greenpeace. "They never came here except for the big hearing, they never showed any interest in working with our group, they never provided facts to back up any of their claims, even when we asked them to, and they never offered any solution to the waste problem other than storage. What kind of a solution is that?" she said.

Greenpeace further alienated itself from Swan Hills residents when, in 1991, one of its officials — Brian Killeen — a marine biologist based in Vancouver, appeared on Edmonton television and was quoted in regional newspapers as saying that he had evidence indicating that a massive epidemic of cancer would spread in the Swan Hills area over the next twenty years because of the waste treatment center. Despite repeated overtures from community leaders and representatives of local environmental groups for further explanation, no data were ever provided to support the claim. Repeated letters, phone calls, and faxes to the biologist all went unanswered. All of this confirmed the initial Swan Hills suspicion of major environmental groups. As one local official explained,

> Some environmentalists like to say we're brainwashed, that we've been duped. There's one man, from Friends of the North, who likes to come to meetings dressed as a coughing vole [a type of mouse common in central Alberta], saying to us, "You must vomit every time you look in a mirror because of what you're doing to the people of Alberta by accepting that facility." We've come to realize

that these people are so angry. They don't want to talk, they just want to get angry. But then there are environmental groups and environmentalists who do care, who want to talk, and who raise good points. FRESH in Swan Hills is a good example. They raise important points and they're listened to.

The public participation procedures put into place during the site selection process were essential in providing a forum for environmental concerns to be raised. Since that time, public access and participation in the facility's operation have remained unusually high. Unlike most waste disposal facilities, the Swan Hills Waste Treatment Centre has maintained a community liaison committee, which is given an office in the small downtown area and some financial support. The committee consists of a variety of citizens not employed by the facility, although Swan Hills is so small that invariably potential conflicts of interest arise. Nonetheless, the liaison committee is generally given high marks for representing local interests. As one former committee chair explained, "The committee serves several purposes. We do serve as a sounding board for the corporation; they try ideas out on us. In turn, we play a watchdog role, and the corporation likes to think of us in this way. Overall, the relationship is good. They've been good, and have to be, because if the trust is ever lost, they have a big problem."

The committee meets at least once a month and convenes more frequently when controversial issues arise. A representative of the corporation also reports regularly at town council meetings and authors a regular column for the local newspaper. When potential contamination issues come up, these mechanisms have worked to swiftly bring debate into the open. For example, facility monitoring discovered some mercury contamination in fish in a lake nearby the facility. This problem was immediately reported to the liaison committee, to the town council, and in the local newspaper. The cause of the contamination subsequently was linked to mercury dumping that had occurred long before the facility was built. A similar process of open exploration of controversies was followed when a leak was discovered in a pipe on the facility grounds, leading to some release of PCBs. The problem was immediately reported to the public and quick remedial action was taken. "I think there's been a real effort to keep the public aware of what is going on," explained a local school official. "Every time there has been any sort of problem, as far as I know, it has been reported by the company and discussed in the open. And the liaison committee has been very honest and capable. They would be quick to catch on and let us know if there was a problem."

Beginning with the commitment to a voluntary siting strategy, Alberta officials have maintained unusual access to the facility during its stages of development, operation, and, more recently, expansion. One of the factors so impressive in this process is the multifaceted nature of the public participation effort in the province. Instead of providing only a few outlets for participation in a specified time period, the province has, with considerable success, devised multiple, ongoing forms of participation. As political scientist William T. Gormley, Jr., noted, drawing from the work of Daniel A. Mazmanian and Jeanne Nienaber, "Public hearings are most effective when combined with less formal and more narrowly focused devices, such as workshops, seminars, citizen committees, and brochures outlining alternative solutions to difficult problems. Public hearings can make a difference when circumstances are propitious and when other participatory strategies are also employed." [18] The multidimensional, longitudinal nature of participation in Alberta hazardous waste management has served as an important cornerstone in reversing the Nimby process so pervasive in the United States and Canada, and so common in Alberta in the years immediately before adoption of this approach. However, voluntarism and public participation are not the only factors that appear necessary to make siting agreements a real possibility.[19]

Institutional Change

Public participation can begin the process of transforming facility siting from a short-lived conflict to a prolonged deliberation over siting and waste management options. But this cannot occur if the basic institutions (public or private) responsible for providing the participatory outlets for consideration of facility siting lack public credibility. Most major players in the private hazardous waste industry in Canada and the United States hardly inspire confidence when they begin siting deliberations, given their controversial track records in prior cases. The public views state and provincial officials with great disdain, attributable in many instances to their ham-handed siting efforts of the past and the general perception that they have been weak in providing basic public protection. States and provinces entered into the hazardous waste regulatory area only recently and have generally failed to earn public confidence. Thus it is difficult to envision existing private or public institutions taking an effective lead role on this issue in the years ahead. The problem is compounded in the United States by the considerable difficulty that the Environmental Protection Agency has had in implementing its Superfund program.

Any legitimate effort to launch serious dialogue on siting and waste management may require basic institutional reform. New, intermediary institutions have often served to provide a fresh start to areas of policy riddled with conflict, both domestically and internationally.[20] Such institutions may be needed to transform the siting process in many provinces and states, distinct from traditional regulatory agencies created in the 1970s and 1980s that lack public credibility and repeatedly fail to attain siting agreements.

The introduction of a crown corporation into provincial hazardous waste management appears to have contributed to the cooperative outcome in Alberta. This corporation assumes a number of important responsibilities delegated to either private firms or regulatory agencies in most states and provinces.[21] It provides for direct governmental oversight of facility operation and also facilitates direct public financial and technical assistance to private corporations responsible for site development and management. As a 1981 provincial report that endorsed the crown corporation concept noted, the corporation

> would provide effective evidence of an arm's length position relative to government and industry . . . while allowing various government departments to continue their particular regulating, inspecting and monitoring functions. . . . The public would be more likely to trust the administration of a crown body. Industry has indicated that if allowed to operate facilities in a free market environment, they too could function efficiently under such administration. Therefore, both concerns are met.[22]

Under this tripartite system — involving the three key partners of public corporation, private corporation, and public regulatory agency — the Alberta Special Waste Management Corporation is responsible for overseeing numerous aspects of the provincial waste management system, including plant design and construction; provision of 40 percent of construction and operating costs plus operating loss subsidies to the private corporation; control of all provincial transfer and collection points; collection of 40 percent of revenues generated by the facility; provision of utilities and highways for the facility; and ongoing research, monitoring, and technological appraisal. It also owns the site, located fifteen kilometers outside Swan Hills, which it leases to the private firm for a minimal fee. Bow Valley Resource Services, a private company, provides 60 percent of the construction funds and operating costs and handles day-to-day operation of the facility.

The crown corporation is distinct from Alberta Environment and related provincial agencies, which set regulatory standards that specify the ways in which wastes are to be treated. Alberta Environment also provides a system for registering wastes and punishing regulatory noncompliance by either the crown or private corporations. At the same time that the Swan Hills siting decision and Bow Valley Resource Services selection were made, Alberta was devising one of the more comprehensive hazardous waste regulatory systems of all the Canadian provinces (or American states), which was to be implemented in part by the crown corporation.

Institutional reform cannot cure all that ails facility siting in Canada and the United States.[23] But the tripartite arrangement does provide a unique distribution of authority and, in the process, eliminates some of the multidepartmental turf battles over hazardous waste that can complicate development of coherent policy. This is a particularly common problem in American states, where multiple departments — including those covering the environment, natural resources, health, commerce, and transportation — are responsible for policy oversight and thereby reduce prospects for an integrated regulatory strategy.[24] This jurisdictional problem occurs in Canadian provinces as well.[25]

An integral, and often overlooked, component of the effectiveness of institutional reforms is the policy professionals — public and private sector officials — selected to implement the programs of new or modified institutions. Their essential tasks include, in the case of facility siting, the ability to foster public trust, develop public dialogue, and pursue other more technical aspects of the siting process.

Hazardous waste facility siting is in many respects a classic example of an intricate form of redistributive policy.[26] Unlike income transfer programs, which shift resources from large numbers of relatively affluent individuals to those who are considerably less affluent, siting involves added complexity and potential controversy. Successful siting confers a broad benefit — a place to send hazardous waste — on a large collectivity of waste generators, but it concentrates costs — responsibilities and risks of facility construction and operation — on a relatively small number of people who live near the facility.[27] Under such circumstances, the role of the policy professional is essential in working within existing institutions to make some form of agreement feasible.

In many states and provinces, the arrogance of policy professionals has helped to undermine siting efforts. While the prospects for a successful siting may have been doomed in any event, they clearly were not assisted by public and private officials who, despite considerable technical expertise, often appear uncomfortable in discussing the matter with the public.[28] Repeatedly, private, state, and provincial officials have demonstrated near incompetence in basic risk communication skills, serving to further undermine market or regulatory siting strategies that were laden with shortcomings from their inception.

The loss of credibility of corporate and Alberta environmental officials — the earlier generation of policy professionals responsible for hazardous waste facility siting — helped undermine the province's market approach of the late 1970s. A 1979 report of the Alberta Environment Research Secretariat indicated that provincial officials were "being seen as aligned with private industry in favour of waste management facilities, as opposed to being neutral." [29] This perception, along with other important factors, served to scuttle facilities proposed for Fort Saskatchewan and Two Hills and ultimately led to the voluntary approach that resulted in the Swan Hills agreement.

The Swan Hills case has brought forth a remarkable coalition of policy professionals from each of the key components of the tripartite system and local government officials. A major conflict did emerge in 1985 over the role of the crown corporation, resulting in the controversial dismissal of the crown corporation chair by the Alberta environment minister. This threatened to return the province to the adversarial days of the late 1970s, although the quick appointment of a highly regarded replacement as chair defused the situation. [30] Important leadership has been provided by a number of key provincial officials with extensive experience in natural resources issues, in both the public and private sectors, and considerable public prominence. Elected Swan Hills officials have provided a solid base of support, with the former mayor and council playing a pivotal role in promoting the project and maintaining a public participation process that could garner trust.

Many observers contend that certain policy professionals representing either Alberta Environment or the crown corporation have been essential to fostering basic trust. In the early stages of public meetings, before site selection, provincial officials Jennifer McQuaid-Cook and Jacquie Champion traveled extensively in the province, meeting with a wide range of citizens and organizations. One Swan Hills official recalled her initial, vehement opposition to the proposed facility, noting that she agreed to attend public meetings to try to block the proposal. But after hearing a full discussion of the nature of the hazardous waste situation in the province, she admitted that she was intrigued. "I had no idea that so much waste was generated in Alberta and that so much of it came from our own town, and our own businesses and homes," she recalled. "It was at this point that I really wanted to know where this stuff was going. I got scared when I learned that no one in Alberta or the rest of Canada really seemed to know."

The meetings began to focus on the possibility of a facility in Swan Hills and what that would entail. The official continued:

> The women running those meetings really knew their stuff. As soon as I didn't understand something, I asked about it, and they put it into language I understood. And if they couldn't answer on the spot, they'd get back to me later with an answer. That made the whole difference. I really felt that these women were above telling me a lie. Everything seemed above board, and they didn't seem to be shoving the plant down our throats. Plus, we knew we weren't the only place that they were looking. That made a difference too. After the meetings, I decided to support the facility. Now, I can say where the waste in my house, my town, and my province goes. Most people in Canada and the U.S. can't say that. Most haven't a clue.

Along with McQuaid-Cook and Champion, Kenneth J. Simpson has played a key role both in the site selection process and in continued management of the crown corporation. He has been involved in hazardous waste management issues through the provincial government since the mid-1970s, serving as senior vice president of the crown corporation before becoming president in the 1990s. He is a familiar figure in Swan Hills and is widely recognized as a trusted leader in the field. The corporation was headed during the mid-1980s by Lorne Mick, a prominent provincial figure with more than two decades of experience in the oil and gas industries. The crown corporation officials have attempted to maintain fairly modest headquarters and a small staff of about fifteen in Edmonton, making regular visits to Swan Hills. They have refused to hire any attorneys, concentrating staff selection instead on policy professionals with expertise in varying physical and social sciences. Simpson, Mick, and the corporation are also responsible for various other aspects of the waste management system, including regional storage and transfer centers, transportation, and efforts to develop a consulting company that would spin off siting strategies and management technologies developed in Alberta to other settings. The corporation has experienced some instability because of loss of personnel, such as Champion, who decided to retire in 1987. Nonetheless, Alberta's hazardous waste policy professionals have been, on the whole, remarkably effective and committed to long-term periods of service, unlike the more transient pattern common in many other state and provincial hazardous waste programs. When combined with the province's unique institutional structure, the prospects for siting via a voluntary system have been enhanced.

Compensation and Protection

Public participation measures and new institutional arrangements can go a long way toward building the sorts of trust necessary for siting agreements to become possible. But a community is highly unlikely to volunteer unilaterally to host a treatment or disposal facility in the absence of tangible economic rewards or clear assurances of long-term safety in facility management. Market approaches to siting have failed in large part because of their near exclusive emphasis on the attractiveness of compensation packages. In the absence of extended dialogue and other trust-building measures, compensation is frequently dismissed as bribery and fuels public animosity toward site proponents. However, as the Alberta case suggests, compensation and related safety guarantees can become a central component to a negotiated agreement when integrated with other essential factors.

The promise of considerable economic benefits from site acceptance was clearly a contributing factor in Swan Hills's willingness to entertain the possibility of hosting a facility. Alberta Environment, the Special Waste Management Corporation, and Bow Valley Resource Services proposed a host of compensatory benefits and safety protections at early stages of the process — and offered them in a concrete manner — instead of waiting for the advanced stages of deliberation over the facility. The economic enticements were particularly appealing for Swan Hills, which suffered a serious recession in the late 1970s and early 1980s as a result of oil and gas price instability.[31] Typical of the province, Swan Hills has historically swung through boom-and-bust periods driven in large part by shifting energy markets.

Observers universally agree that the facility has contributed substantially to a direct turnaround of the local economy. "It has clearly stabilized our town," noted one resident. "Given the depletion of the oil fields and layoffs, we might be a ghost town by now were it not for the plant." A 1991 report concluded that the facility provides more than ninety full-time jobs in Swan Hills that contribute $2.7 million a year to the local economy in salaries and makes an overall impact of $6 million on the economy each year. Total personal income, trading area income, and retail trade have grown at rates well above inflation or the overall Canadian economy between 1978 and 1988.[32] In 1986 Swan Hills's unemployment rate was 7 percent, among the lowest in the province, and its average household income, $44,023, was more than $7,000 above the provincial average.

Housing, a long-term problem because of population transiency and traditional reliance on mobile trailers, has stabilized somewhat through corporation construction or renovation of nearly three dozen single-family dwellings, which are usually rented to facility employees. New housing starts and building permits also took a significant jump; nearly sixty new homes were built between 1987 and 1990, contributing to the sense that Swan Hills is beginning to become a community of neighborhoods instead of transient rental housing and mobile parks. Home Oil Company has built a 10,000-square-foot office complex to be shared with other firms and has also entered the housing market. Some steps have been taken to develop an industrial park, with the intent of drawing firms eager to be located near the waste treatment and disposal facility.

The facility also may lure training programs for environmental technicians and lead to export of locally developed waste management technologies. "There's still a perception around some parts of Alberta that we're some big garbage dump," lamented a town councilor. "But we respond by saying that we are becoming the environmental capital of North America. We can say to other communities and provinces that we're in the middle of the new industrial revolution, trying to find ways to clean up all the things you people screwed up. We're the future. Besides, if it wasn't for us, where would all these other communities be, what would they do with their waste?" Provincial efforts to develop training programs and export waste technologies remain in the early stages of development, however; the role they will play in long-term economic development remains unclear.

Swan Hills has also proven adept at securing government and corporation support as part of its overall compensation package. The town received grants for development of a golf course and other recreational facilities, special rescue equipment for the fire department, supplemental materials for the local schools, hockey schools, 400 trees planted around the community, and a bear rug for town council chambers. The town was also successful in securing a modern, twenty-five-bed, $7 million hospital and a $5 million upgrade of water supply facilities, although these cannot be linked formally to its decision to accept the facility. An unexpected economic benefit is tourism, which was virtually nonexistent before the facility opened. Local officials estimate that 3,000 visitors come to Swan Hills each year, some to take advantage of skiing, golfing, hunting, and related recreational opportunities and many others to visit or study at the facility. As a result, the town that seemed to have such an uncertain future little more than a decade ago now has four motels and counts tourism as a modest but steady supplement to its economic base.

Local residents contend that the most important aspect of the economic expansion has been its stabilization. "The kinds of jobs that were dominant here before were very unreliable," noted one veteran school official. "People would be here a few years and then they would lose everything. These families had it very rough and we'd see kids with lots of problems. These problems have hardly disappeared but what we're seeing are more stable jobs and families, allowing us to become a much more stable community." Other citizens agree and contend that the facility may serve to promote public health. Despite any potential dangers related to waste management, citizens note that economic stabilization appears to have led to a reduction in domestic violence and episodes of alcohol and drug abuse, although no formal evidence is available to substantiate this claim.

Perhaps the biggest single controversy concerning the long-term social and economic future of Swan Hills has nothing to do with the waste treatment and disposal facility. The town has a kindergarten-through-eighth grade school, which is well equipped and generously endowed in per pupil expenditures compared with other Alberta districts. However, the nearest high school is ninety-seven kilometers southeast in Barrhead. This requires a grueling commute that makes long-term residential commitment to Swan Hills unattractive to parents of teenagers. Residents concur that the building of their own high school, along with continued safe management of the plant, will be essential to the long-term social and economic vitality of Swan Hills.

Economic support, however, has not been the only part of the package promised to Swan Hills in exchange for its acceptance of the waste facility. The province has established a fund, held by an independent third party, to assure perpetual care of the landfill site following its eventual closure and to cover twenty years of monitoring. The crown corporation also agreed to many revisions in its original plans to accommodate local concerns and suggestions. These included expanded stations for monitoring air and water quality around the plant, a unique system of grids to tighten the monitoring process, construction of buildings to shelter open landfills before their closure, and movement of the facility to a basin location to avoid possible runoff damages. An agreement also was reached to prohibit any deliveries of wastes to the facility after dark, in response to complaints from Swan Hills citizens about the noise generated by trucks moving through town at night. The corporation has remained open to suggestions for further arrangements as desired by the liaison committee, town council, or general citizenry.

The compensation process has not, however, been without flaws. First, many Swan Hills officials acknowledge that they committed a fundamental error in not securing from the corporation regular fees or tax payments for facility operations. Because the facility is located outside town boundaries, such revenues go to a regional development authority, not Swan Hills, making facility acceptance far less lucrative for the immediate local government than it might otherwise have been. Second, periodic charges of favoritism and conflict of interest have surfaced in the dispensing of rewards and jobs. Some local leaders received free trips to Europe to visit comparable facilities before the siting decision was made; a former mayor became a paid consultant to the corporation after leaving office; a town councilor agreed to serve as a paid member of the corporation board; and a few early site opponents were ultimately offered jobs at the facility. Such examples do raise some question of propriety. However, few observers view them as anything other than mild forms of co-optation, none of which violate any federal or provincial laws or deviate from traditional industrial siting practices. A former chair of the liaison committee explained that "as a liaison group we try to keep away politically. Some people think we're bought off but that's just not the case, even though there are some politics involved. Last year, I put in an enormous amount of time on the committee, even though I work full-time as a school teacher and don't get paid for being on the committee. My rewards? I got four tickets to a hockey game, a plaque expressing appreciation for my efforts, a free dinner and chocolates at Christmastime, and free refreshments during meetings. That's all."

Even those critical of the facility, such as representatives of environmental groups, concur that the compensation packages have been largely free of corruption and have greatly benefited Swan Hills in economic and social terms. When compared with the far-reaching charges of deception and coercion common in efforts to site waste treatment and disposal facilities elsewhere, any concerns raised by these benefits are minor. "We entered this process on the assumption that the corporation lies to us," noted the coordinator of Edmonton-based environmental groups. "But what's emerging in the process is that they have done credible work." Moreover, the combination of economic compensation and safety guarantees, when integrated with the other features of the Alberta siting approach, was essential to securing the publicly supported agreement.

Sharing the Waste Management Burden

Hazardous waste facility siting under either market or regulatory approaches usually involves no linkage to other aspects of waste management that would more broadly distribute the burdens borne by various communities. With its singular focus on getting some community to accept a site, no integration is made with other features of a waste management system that might reassure a potential host community that it is not about to assume near-exclusive responsibility for waste generated over a wide, perhaps unpredictable, geographic area. Targeted communities are understandably fearful that, should they agree to host a facility, they may become a magnet for wastes shipped from long distances (and multiple states and provinces unable to develop their own capacity). They also fear that facility acceptance may undermine the willingness of other communities to house their own facilities, even for waste storage, transfer, and recycling. Finally, they fear that if a facility is opened in their community, incentives for industries and communities to recycle and reduce their wastes will diminish or disappear, because of the greater availability of treatment and disposal. All these reasons, combined with the numerous other incentives to reject market and regulatory siting strategies, contribute to the regular defeat of siting proposals.

Alberta's successful siting strategy addressed each of these legitimate concerns, integrating siting of a comprehensive treatment and disposal facility with a larger set of commitments to burden sharing for waste management. As a result, communities such as Swan Hills, Ryley, Veteran, and others that expressed interest in hosting a facility were reassured that a central part of the siting process involved special commitments to distribute the burden for waste management more fairly than under the sorts of narrowly focused siting efforts conducted unsuccessfully in places such as Fort Saskatchewan, Two Hills, and many other communities in Canada and the United States.

IMPORT CONTROL. Restricting the movement of wastes, including hazardous ones, across governmental borders is an extremely sensitive issue throughout the world. Many, including the U.S. Supreme Court, contend that hazardous waste is a good no different from television sets and widgets and that states cannot restrict the interstate movement of these goods based on the commerce clause of the Constitution. Many states and localities continue to fight this interpretation, however, and federal legislation has been proposed to give states and regions greater control over their borders. Moreover, American governmental entities that play a direct role in constructing and operating waste facilities, instead of merely overseeing privately held ones, are thought to possess restrictive powers under well-established "market participant" doctrines. But in Canada, as in much of Europe, fewer impediments exist, as individual provinces have considerable latitude to control the movement of wastes. Alberta has used this power aggressively in building support for the facility eventually opened in Swan Hills.

A central component in the public participation process in the early 1980s was a discussion of how reliant Alberta had become on waste exports, particularly to the United States. As one provincial official noted, industries were sending about 1,000 cubic feet of PCBs to a facility in Oregon, several hundred kilometers away, each month. "I feel it's irresponsible for us to send our wastes across the border," he said in a public hearing. "And I would be pleased to see the [U.S. Environmental Protection Agency] stop the practice." [33] His suggested alternative was to end exporting and promote provincial self-reliance for waste management, which would include the construction of a comprehensive facility and waste management program.

At the same time provincial officials lamented the reliance on waste exports, they also emphasized the need to restrict any facility that would be developed within the province to provincially generated wastes alone. A 1980 Environment Canada report called for the construction of a single, massive hazardous waste facility to be located in southern Alberta that would serve the needs of all western and central Canadian waste generators. This proposal was immediately rejected by the Alberta environment minister, who said, "We were not intending on becoming a dumping ground" for other provinces, and the chair of the Environment Council of Alberta, who termed the proposal "absolutely daft." [34]

Instead, Alberta officials consistently advanced the proposition that any comprehensive treatment and disposal facility would be intended exclusively for Alberta wastes. As Alberta Environment research scientist Natalia M. Krawetz, who played a key role in developing the voluntary approach, wrote in a 1979 report, "In general, people tend to accept the need for a waste treatment facility if it appears to be solving a local problem; but there's a reluctance to treat wastes generated elsewhere." [35] The lone exceptions to the ban on imports, it was stated, would be formally arranged reciprocal agreements, whereby Alberta and one or more other provinces or states might share waste management responsibilities depending upon their particular form of treatment or disposal capacity.

Most observers concur that this commitment to restrict imports played a significant role in convincing Swan Hills residents to accept the comprehensive facility. "Import control was a central part of the policy from the start and has been a key part of the process here," explained one resident, who formerly served on the liaison board. Another citizen with a keen interest in the facility represented the views of several interviewees:

> I think a lot of people have been open to the possibility of taking wastes from places like the Yukon and Northwest Territories. They're poor, small, and don't

generate a lot of wastes. And even Saskatchewan would be a possibility. They're a hard luck province; they probably couldn't afford a facility of their own and they don't generate much waste anyway. But forget it as far as the other provinces are concerned. BC [British Columbia] is rich, and it's up to them to take responsibility for their own wastes. I don't want to see them shipping their wastes across the mountains to Swan Hills, and I know a lot of people feel the same way. The same goes for eastern Canada, especially Ontario and Quebec which generate so much waste. Let them take care of their own problems. And I'd say absolutely no to any U.S. wastes; they have plenty of places to put them and plenty of money to pay for it.

Since the opening of the plant in 1987, Alberta has neither imported nor exported virtually any hazardous waste, the only state or province in the United States or Canada that can make that statement. One proposed exception was an Albertan offer to the government of Quebec to take PCB-laden wastes generated by the 1988 warehouse fire in St.-Basile-le-Grande. However, as one liaison committee member recalled, "The corporation came to us and asked what we should do, and we agreed to take the waste on a humanitarian basis. Quebec had no proper facilities and it was intended as a one-time-only deal, with Quebec picking up all the costs. But then we ran into problems. Quebec refused to ship the waste piece-by-piece and wouldn't re-pack it as we thought appropriate for shipment across the country. So we told them to forget it. Then they had a real mess, sending it to the United Kingdom, where it was rejected. Finally, it wound up being stored in, of all places [Prime Minister Brian] Mulroney's riding [a legislative district within a Canadian province] in [Beie Comeau,] Quebec." In 1985 Alberta elected to expand its ban on imports to the Swan Hills facility to any other hazardous wastes that might be shipped to a small provincial storage or treatment facility. This decision was prompted by a spill along the Trans-Canada Highway in northwestern Ontario of hundreds of liters of PCB-contaminated transformer oil from a private truck, which was headed for a then-privately owned storage facility in Alberta.[36]

These episodes rigidified Alberta's resistance to hazardous waste imports. However, neighboring provinces, particularly British Columbia and Saskatchewan, have remained particularly eager to gain some access to the facility, given their own siting inabilities. A potential controversy looms as Alberta pursues expansion of the facility. The proposed expansion would add considerable incineration and stabilization capacity, and provincial officials have begun to explore the option of accepting out-of-province wastes, at least on a formally structured basis. Officials have characterized this as a shift from a "province-only" to a "province-first" approach, although non-Canadian wastes would not be accepted under any circumstances. Public consultation on the option began in November 1993.[37] Driving forces behind this possible shift in policy have been the reluctance of the province to continue to subsidize the facility or increase charges for Alberta users and the possibility of drawing significant revenues from wastes from other provinces desperate for sophisticated treatment and disposal technologies.

Public hearings on the expansion in 1991 and 1992 drew little controversy but suggested that out-of-province wastes would not be accepted.[38] The official environmental assessment report makes a strong case for the expansion, largely on the basis of needing to accelerate the cleanup of extant wastes, but dodges the issue of whether exports will be allowed to enable the corporation to make full use of its facility.[39] Interviews during those years indicated mixed viewpoints on the subject of accepting imported wastes. Officials from the crown corporation and town government are generally supportive of limited import acceptance, while members of the liaison committee and other prominent citizens are somewhat skeptical, especially if large quantities are involved. Opponents contend that any perception that Swan Hills was becoming a magnet for non-Alberta wastes or any episode of controversy caused by transport or treatment of wastes could serve to undermine the strong basis of support that has endured for nearly a decade. They also emphasize the issues of long-distance transport and possible widespread dependence on Swan Hills for many years to come. In any event, the commitment to restrict the facility to Albertan wastes contributed significantly to the atmosphere of trust surrounding siting deliberations in the mid-1980s, contrary to the more typical pattern whereby Nimby groups understandably hold out the likelihood of becoming regional waste magnets as a further reason for opposition.

REGIONAL STORAGE FACILITIES. Restrictions on waste imports were not the only provision intended to promote more equitable burden sharing in the Albertan siting process. Officials acknowledged that the brunt of waste management would be borne by the community hosting the comprehensive facility but emphasized that other communities would be expected to play a direct, albeit supplemental, role. The corporation agreed to develop a regional system of waste collection and storage, before transfer to Swan Hills. These facilities were intended to serve a number of possible functions, including receiving, content analysis, unloading, segregation and handling, interim storage, repacking, bulking, pretreatment, and reloading for final shipment. At about the time the central facility was becoming operational, the province opened a waste storage station in Nisku, a suburb located five miles south of

Edmonton.[40] Ryley followed its receptivity to the provincial siting initiative —
and its anger over being rejected — by contracting with Newalta Corporation
to establish a waste storage station and a small landfill for treated inorganic
wastes.[41] The large urban center of Calgary has also agreed to host a waste
storage station. Both the Nisku and Calgary facilities were already providing
some waste management services, making public acceptance easier. Additional storage stations are contemplated in future years, along with an expansion of neighborhood collection programs.

The Nisku and Calgary agreements were reached under voluntary provisions similar to those used for the waste management center in Alberta. Other
communities have also expressed interest in hosting waste facilities, although
in at least one instance, involving the Edmonton suburb of Morinville, failure
to inform the community fully of the types of materials that would be handled
led to a strong public outcry and withdrawal from the process.[42] This experience has only bolstered provincial commitment to the open, voluntary process
in future storage facility siting.

The key linkage between these various pieces of the waste management
system was a unique approach to waste transport by truck. In most states and
provinces, waste transport is conducted through a variety of private firms, with
little, if any, special designation of vehicles carrying wastes or uniformity of
design. Hence citizens have little sense of what vehicles carry hazardous waste
or where the cargos are headed. In Alberta, crown corporation officials sought
a more secure, predictable waste transportation system. Specially designed
and marked vehicles, including tractor trailers, bulk tankers, specialized flat
beds (for hauling transformers, which have proven ubiquitous in Alberta), and
dump trucks (for contaminated soils), comprise a fleet operated by the corporation. Virtually all wastes shipped around the province, and to Swan Hills for
final management, must be conveyed in standardized — and uniformly
marked — vehicles.[43] When combined with the regional storage points, local
collection stations, and the central facility in Swan Hills, this gives the province a unified system that contrasts markedly with the largely unknown web
of waste movements in other states and provinces, adding further to the sense
of trust in the system.

WASTE RECYCLING AND REDUCTION. A final component in the burden-sharing strategy involved multiple commitments to increase public awareness of
the widespread generation of hazardous waste in the province and communal
responsibility for safe disposal and recycling or reduction where possible. Alberta moved into this area more rapidly and extensively than most other prov-

inces and, in many respects, rivals the more innovative states, such as Minnesota, in the extent of its commitment to waste recycling and reduction as a component of an overall waste management strategy. Some corporations and communities have moved aggressively into these areas. For example, Syncrude Canada, which operates the world's largest synthetic crude oil production facility, forty kilometers north of Fort McMurray, has introduced an across-the-board approach to reduction and recycling. With provincial assistance in some instances, employees have found creative ways to switch to various biodegradable chemicals and reuse hydrotreater and other catalysts, glycol, and numerous metals utilized in manufacturing, such as nickel, zinc, copper, and silver.[44]

Many recycling and waste reduction efforts were launched in the mid-1980s and were related to the desire to locate a comprehensive disposal facility. The Alberta Special Waste Services Association was created with the charge of developing household hazardous waste collection programs, establishing informational materials, and encouraging residents and merchants to bring hazardous materials to a safe storage facility, where they could be shipped to Swan Hills for treatment or disposal. The association complements the Alberta Waste Materials Exchange program. Established in 1984, it provides a clearinghouse for products that for some may be waste but for others may be productively used.

Specialized programs have been established to allow convenient drop-off of a number of hazardous materials for recycling or reuse. More than 250 depots have been created for collection of agricultural chemical containers; in 1992 more than 785,000 containers were retrieved and recycled, with 19,000 liters of hazardous waste from the containers sent to the center for disposal. Alberta has also established an oil drop program that offered consumers 350 drop-off points in service stations, from which waste oil can be re-refined and containers recycled. Approximately 2.2 million liters of used oil were recovered through this program in 1992. Similar programs have been established for products such as dated pharmaceuticals, which may not always formally meet the technical definition of hazardous waste but nonetheless pose a public health threat if disposed of improperly. In 1992, twenty-nine tons of old, unused, or outdated drugs were gathered.[45] Used refrigerators and freezers have also been targeted in such a manner, with their chlorofluorocarbons (CFCs) drained and recycled.

Provincial officials have also attempted to promote recycling by easing certain regulatory provisions for hazardous materials that can be recycled or somehow reused instead of sent to Swan Hills for final disposal. Moreover, the recycling and reduction activities have been supported by an active campaign of public education and information. The provincial crown corporation has developed an extensive K-12 curriculum that explains hazardous waste and waste management practices and also promotes waste reduction and recycling. "All of this is designed to make people stakeholders for the wastes they create," explained one local official in Swan Hills who is familiar with these efforts. "The educational programs are taken very seriously. Even here in Swan Hills, where we have all of this disposal capacity, kids are laughed out of class if they bring a sandwich to school wrapped in plastic wrap rather than something reusable." Provincially based schools have also been the focus of cleanup efforts; a two-year program throughout the province led to the discovery and removal of 4,000 205-liter drums filled with hazardous waste.[46]

Officials responsible for development of these efforts contend that the simultaneous emphasis of waste recycling and reduction alongside facility siting is essential for the political and operational success of a comprehensive waste management system. "We had somewhat of a head start in Alberta, at least with solid waste recycling," recalled a senior provincial official. "We developed a blue box recycling program for paper and glass and placed a compulsory deposit on beverage containers in 1972. That paved the way for some of the things we did in hazardous waste in the 1980s. What we've said all along is that waste recycling and minimization have to coincide with the other parts of the system. But we also contend that there will always be residues and wastes that require treatment and disposal."

Perhaps the biggest limitation of the Alberta approach to waste reduction and recycling is the lack of clear mandates and incentives for major waste generators to pursue these strategies. Because of the availability of the Swan Hills facility and the provincial subsidization of costs to waste generators that use the facility, some contend that reliance on treatment and disposal has become too easy an option. Such debate is difficult to analyze fully, for Alberta, much like other provinces and states, has not yet developed fully reliable measures of the volumes of waste reduction and recycling that are taking place. Nonetheless, Alberta remains one of the few provinces to pursue recycling and reduction systematically and one of very few provinces or states to have fully integrated them into a comprehensive waste management strategy.

System Performance

By any reasonable measure, Alberta has made enormous strides in developing a comprehensive system of waste management. On both technical and political grounds, recent developments in the province are impressive. Alberta reversed the Nimby pattern so pervasive in Canada and the United States with an approach that combined voluntarism and a multifaceted system of protections and incentives for communities that would contemplate the possibility of hosting a site. Traditional patterns of haphazard waste dumping have been replaced with a sophisticated system that includes storage, treatment, disposal, reduction, and recycling, all of which operate under a broad base of political support. This base remains solid in the community hosting the comprehensive facility, more than a decade after it chose to volunteer to host the facility. By the standards of most states and provinces, this is truly remarkable. The Alberta case, however, is neither free of shortcomings nor a sure recipe for success elsewhere.

Swan Hills as a Fluke

Some analysts dismiss the Swan Hills case, contending that a combination of its economic status and unusually isolated location contributed to a siting success that is unlikely to recur. Others maintain that Alberta has a distinctive political culture, which is highly individualistic and amenable to taking risks. These arguments have some merit but cannot be taken too seriously. Numerous other communities in Alberta, as well as other provinces and states, feature many of the same qualities as Swan Hills. Many suffered significant economic problems in the 1980s, some more serious than those faced by Swan Hills. Furthermore, many communities exist with few nearby neighbors. A number of comparisons can be drawn, both economically and in the physical isolation of individual communities, between Swan Hills and areas such as the vast Upper Peninsula of Michigan. Many communities have been targeted for facilities under market or regulatory approaches, but they have not proven easy marks for siting proponents. Moreover, Swan Hills was not an isolated case within Alberta. Multiple communities within the province came forward as potential site volunteers, none featuring economic or geographic characteristics that were identical to Swan Hills. Additional communities, including large cities such as Calgary and a suburb just outside Edmonton, have voluntarily accepted regional transfer facilities. As to the issue of political culture and whether Albertans are somehow less risk-averse, it should be noted that three prior siting efforts failed in the province. In addition, the siting of solid waste landfills, handled through traditional siting methods in Alberta, has been mired in Nimby-like controversy for years. Unlike the provisions the province has made for hazardous waste, those that it has made for solid waste have resulted in a serious shortage of disposal capacity. What appears to have made the difference in the Alberta hazardous waste case is neither unique economic, logistical, or political culture factors but a radically different approach to siting that shows some signs of being replicable elsewhere.

Unanticipated Costs

Perhaps the biggest single concern over the continued viability of the Albertan waste management system, assuming continuation of safe management practices, is unexpectedly high costs of operation. Alberta committed a fundamental error in deciding on the scope of its treatment and disposal facility before it had reliable estimates on the volumes and types of wastes that were being generated and had been abandoned by prior generations. Given the joint nature of the crown-private corporation partnership, the former has retained responsibility for subsidies. These have continued to range from $15 million to $20 million a year, although officials say they are likely to decline in future years. Because of the poor capacity planning, some parts of the facility have been underutilized, whereas others are inadequate, given demand, and may require capacity expansion. This has triggered provincial support for the $60 million expansion of facility capacity, partly because of the large backlog of organic wastes that were abandoned around the province and require incineration. A major component in this expansion is a new furnace specially designed to decontaminate metal in electrical transformers, tens of thousands of which have been discovered in varying states of deterioration around the province. The expansion also involves creation of a stabilization facility to solidify sludges and solid waste residues before landfill.

Concern about the ability of the province to continue to provide subsidies has mounted. Alberta has faced particularly serious budgetary shortfalls in 1993 and 1994, although it has left its waste management program largely untouched. Looking ahead, officials feel they have two choices if pressured to eliminate subsidies. First, acceptance of hazardous waste imports, even in restricted amounts, could greatly alleviate the need for further subsidies. Neighboring provinces would probably be willing to pay at rates well above what Alberta charges within-province generators. Unlike states, provinces should experience no constitutional barriers to setting differential waste disposal rates according to locale of generation. However, this strategy would represent a direct break of the original commitment to restrict waste imports and could open a series of political and environmental problems. Second, Alberta could choose to charge waste generators a good deal more for their use of the Swan Hills facility than they are at present. To date, officials have been reluctant to do so. They view provision of hazardous waste treatment and disposal as part of the infrastructure governments should provide if industry is to thrive. They are reluctant to raise rates significantly, for fear that generators might return to the more slipshod disposal methods of the past. Officials may also be chary to adopt such a strategy because industries could move their operations elsewhere, although this outcome would seem unlikely given the consistent research finding that environmental regulatory rigor has little impact on industrial location and relocation.[47]

Although the subsidies are a source of concern and may leave the province with a difficult policy choice, hazardous waste management, by the standards of most states and many provinces, continues to be conducted inexpensively in Alberta. Not only can the province assure existing or incoming industry of a sophisticated, reliable treatment and disposal system, but it can also assure a fairly stable level of total costs related to waste management. Treatment and disposal of wastes abandoned in prior generations pose little of the explosive legal and technical costs borne elsewhere, both in the United States and increasingly in parts of Canada, because of access to the central facility managed by the province. Moreover, the costs of the siting process itself were low in Alberta. Whereas Ontario has spent more than $100 million for a process to search for a comprehensive facility with little to show for it, Alberta spent far less than that to complete its siting process and facility construction.

In comparison with expenditures by American states, the costs of Alberta's waste treatment and disposal remain a bargain. For example, states such as New Jersey have devoted hundreds of millions of dollars to hazardous waste management, nearly one-half of their total environmental expenditures in recent years, with overall results strikingly similar to those of Ontario.[48] Thus, multimillion-dollar provincial subsidies are understandably unpleasant to swallow and may reflect some poor planning, but they are a drop in the bucket compared with the overall costs of hazardous waste management in other parts of Canada and all of the United States.

4

Prospects for Replication

A SINGLE SITING AGREEMENT does not constitute an airtight case for the voluntary, comprehensive approach to hazardous waste facility siting. Although the Alberta agreement cannot be dismissed as a fluke on the basis of any unique characteristics, the larger question remains as to whether its basic tenets can be successfully applied in other settings in Canada and the United States. The relative newness of the Alberta approach and the continuing emphasis on traditional market and regulatory strategies in most provinces and states have served to preclude any wide-scale testing of it. Nonetheless, a small but growing number of examples indicate serious prospects for replicability. In some cases, the siting efforts were modeled directly on the Alberta experience. In others, common features were developed in a coincidental fashion. Swan Hills–like siting agreements and the construction of a comprehensive waste management system, as in Alberta, were not the only outcomes of these efforts. But in many respects, they point to a reasonable expectation that within the Alberta approach, and its design characteristics, lies the beginning of a more promising means to siting than those that continue to dominate in Canada and the United States.

Emergence of the familiar Nimby outcome need not be an inevitability. Provisions for a provincewide facility for Manitoba and a smaller one for Greensboro, North Carolina, offer models of voluntary, comprehensive siting approaches that result in siting agreements. Related cases in Minnesota, California, and Quebec suffered shortcomings but indicated the potential for state and provincial officials to make progress toward creation of public dialogue on facility siting and ultimate development of a more integrated — if not fully comprehensive — system of waste management.

Siting proved more successful when provinces or states used the combined forces of the voluntary approach and assurance of wide burden sharing for waste management. As in Alberta, a voluntary process was established in Manitoba and North Carolina that promised considerable protections to potential host communities, including continuous public participation through the stage of facility management, adherence to safe management practices, and measures to protect new facilities from becoming magnets for massive quantities of wastes from outside the immediate area, state, or province. Strong emphasis was placed on public participation and involvement throughout the siting process, not the top-down style of siting selection common in market and regulatory strategies.

The Minnesota, California, and Quebec cases were somewhat less successful in these respects. In Minnesota, the institutions responsible for fostering dialogue over siting and waste management became mired in controversy at latter stages of the voluntary siting process and lost any capacity to earn public trust. In California, a promising approach emphasizing formal burden sharing among counties was undermined by state agency opposition. And, in the case of Quebec, post-siting failure to adhere to assurance commitments has weakened public support for the facility that was opened near Montreal with strong local approval in the early 1980s. Nonetheless, all of these cases went beyond the traditional siting pattern and outcome. They thereby further confirm the potential for the voluntary, comprehensive approach to siting.

Manitoba: Alberta Redux

The diffusion of policy innovation across subfederal boundaries is well understood in the United States, thanks to pioneering research efforts of the 1960s and 1970s.[1] While the process has been less fully studied in Canada, application of the Alberta siting to Manitoba offers a classic case study of how policy innovation — and success — in one province can prompt a neighboring province to rethink and revise its policy. Not only did Manitoba borrow heavily from the Alberta experience in devising an approach to hazardous waste facility siting and management, but it also experienced an almost identical outcome. After an extensive period of public deliberation, multiple communities came forward as volunteers for a comprehensive facility, three of which proved particularly interested. In September 1992 ground was broken for construction of a publicly accepted comprehensive facility in the Rural Municipality (RM) of Montcalm. This facility, known as the Manitoba Environmental Centre, was expected to open in 1994.

Much like Alberta and other provinces and states, Manitoba assumed an extremely lax regulatory stance toward hazardous waste before the 1980s. This central Canadian province has 1.1 million citizens and generates approximately 180,000 metric tons of hazardous waste a year. It has never had any sophisticated treatment or disposal facilities within its boundaries. A 1989 provincial report confirmed the view of most observers that "the largest proportion of waste is presently being sewered." [2] Other common techniques included landfilling and on-site storage and, to a lesser extent, export outside the province. In many communities, direct dumping of wastes into major waterways took place; this was a particularly serious problem for the Red and Assiniboine rivers that pass through Winnipeg, the capital and largest city of Manitoba, which generates more than half of all hazardous waste produced in the province.[3] Despite the dominance of Winnipeg industry in provincial waste generation, hazardous wastes are produced all around the province, by agriculture, secondary and service industries, as well as industrial mainstays, making it more than a problem confinable to a single, large municipality.[4]

The province began to examine proper waste management seriously in the early 1980s, attempting to develop regulations. But as one veteran provincial official noted, "It is especially difficult to enforce them when there is no alternative such as a facility to manage the waste." Officials were unsuccessful in attracting private waste management firms to propose treatment or disposal facilities. They held out some hope of access to the proposed Alberta facility, if its voluntary siting process were to prove successful, but recognized that it was being designed to block out-of-province efforts.

Instead of trying to pry their way into the new Alberta facility, Manitoba officials chose to study the successful process that the province used in the mid-1980s. The Alberta approach thus became a model for many aspects of what Manitoba would do, with modifications where appropriate, given the differing economic and political circumstances facing the two provinces. Much like Alberta, Manitoba developed a crown corporation and gave it responsibility for implementation of the public participation program. In the first years of the process, provincial officials held hundreds of meetings with local government councils and various community groups, such as public interest groups, industry associations, and service organizations. Dozens of additional meetings were held with individual communities at advanced stages of the process.[5] Much like the situation in Alberta, Manitoba officials generally found smaller open houses and public meetings to be more effective than large public hearings in which only a small fraction of participants were able to speak. As a result of the meetings, more than forty communities expressed interest in the possibility of hosting a comprehensive treatment and disposal facility.[6]

Many of these communities either withdrew from further consideration for political reasons or were asked by provincial officials to drop out of the running because of potential environmental problems concerning their locale. Manitoba remained in the unique position of having five communities active in the pursuit of the facility as of 1990: the heavily populated capital of Winnipeg, as well as the smaller communities of Hamiota, Montcalm, Pinawa, and Rossburn. Some political conflicts emerged in Hamiota and Rossburn, and referenda on the facility were narrowly defeated in both communities in 1990.[7]

With three actively engaged communities left, Montcalm jumped to the top of the list after a September 25, 1991, referendum was approved: 67.1 percent of voters supported the facility proposal and voter turnout was approximately 75 percent. Pinawa had indicated strong support and offered the advantage of having a highly skilled work force nearby, as it is the site of the Whiteshell Nuclear Research Establishment.[8] A number of technical concerns about site suitability made it less desirable than Montcalm, although it subsequently became the center for developing, installing, and operating a comprehensive monitoring system for the Montcalm-based facility. Winnipeg was attractive because much of the province's waste would be located within easy reach of the proposed facility. Moreover, given that Winnipeg generated so much of the overall waste, some argued that it should bear the greatest responsibility for waste management.[9] However, some political opposition began to surface from local environmental groups and a few local members of the provincial parliament.

As a result, Montcalm emerged as the finalist; it was an excellent candidate according to technical criteria, and in the words of one provincial official, "It was hard to argue with an overwhelming referendum vote." Once the announcement was made, local reactions were "generally jubilant," more reminiscent of the acquisition of a professional sports franchise than a hazardous waste facility. The community's representative in the provincial legislature declared, "This is a recognition of the tremendous effort put out by the people in the RM of Montcalm and by the council of Montcalm." [10] Many Winnipeg officials expressed their displeasure at being bypassed in favor of Montcalm. Mayor Bill Norrie said, "I'm terribly disappointed." His views were shared by members of the Winnipeg advisory committee, who argued that the central facility belonged in the province's largest city to minimize transportation risks. Project opponents, in turn, expressed relief.[11]

The Manitoba Environmental Centre will be on a sixty-four-hectare site located approximately seventy-five kilometers south of Winnipeg, about seven kilometers outside the villages of St. Jean-Baptiste, St. Joseph, and Letellier, which combine to create the Rural Municipality of Montcalm and its 1,700 residents. The three villages participated in the referendum, and informal evidence indicates strong support for the facility from outlying communities as well.

Like the Swan Hills facility, the one under construction in Montcalm will serve as a central treatment and disposal facility for all hazardous wastes generated in the province. Provincial officials recognize, however, a smaller and less diverse body of hazardous wastes than their western neighbor and have accordingly planned for a somewhat smaller and different facility. The center will have capacity to treat an estimated 11,000 tons of inorganic waste and 5,000 tons of organic waste per year. In November 1993 the corporation opened a separate component of the facility, which can treat up to 40,000 tons of petroleum-contaminated soil per year. It uses biological processes to break down organic contaminants, primarily from leaking underground fuel storage tanks. Not only will the Manitoba facility be physically smaller than its Alberta counterpart, but it will also refrain from operating an incinerator of any sort and will, in general, place greater emphasis on treatment than disposal technologies. This reflects a philosophical difference to minimize reliance on the technologies thought to pose the greatest environmental and public health risks and also tailor the facilities to the needs of Manitoba hazardous waste generators. Despite these important differences, Manitoba adhered to the four design characteristics of successful siting demonstrated in the Alberta case: extensive public participation; institutional change to win public respect and trust; a package of economic compensation and safety assurances; and distributing the waste burden equitably with other Manitoba communities.

Public Participation

Manitoba was more adamant about building public participation into every dimension of the siting process than was Alberta. Provincial policy documents repeatedly reflect the sentiment expressed in a 1988 public report that "it is essential that the process be flexible, evolutionary and incorporate a proactive public and stakeholder participation component." [12] A prominent provincial official confirmed this emphasis, noting that Manitoba used Alberta's open approach as a model to be built upon and Ontario's top-down approach as a model to be avoided. "We have recognized that public support is essential, the only way we're going to be successful," he stated. "We're not just paying lip service to participation."

The participatory process in Manitoba included a barrage of educational materials, ranging from brochures and displays at shopping malls to sponsorship of videos filmed by the province's educational television station. The crown corporation sponsored more than 500 committee and public meetings, open houses, and presentations. Moreover, provincial officials combed virtually every corner of the province where siting was technically conceivable. This extensive involvement only intensified as communities began deliberations over whether to declare themselves candidates for the site. Even in Rossburn, which voted against remaining a candidate by a 376-to-279 margin in 1990 and experienced some serious community tensions over the issue, numerous public meetings were held, a number of educational and participatory opportunities were provided, and a solid base of support for the facility endured.

In Montcalm, turnout was relatively light at a pair of open houses held in November 1989. The provincial crown corporation responded by holding an additional pair of open houses in February 1990; it sought to bolster attendance by hiring high school students to contact every Montcalm household and personally invite all residents to participate. Such an outreach effort is virtually inconceivable under traditional market or regulatory approaches, where absence of public involvement would be seen as a good omen for imposing a facility with little discussion. For this second round of open houses, 250 residents attended and 84 percent of the 186 questionnaires returned by the citizens who participated in the meeting supported additional study of a facility for Montcalm. [13]

Deliberations continued, and a local advisory committee was formed. As in Swan Hills, the crown corporation made some effort to win the support of committee members through tours of other Canadian and American waste management facilities. Some opposition to the proposal began to surface in late 1990, led by the Concerned Citizens for Ecology and Environment. This group presented petitions with 450 signatures, constituting more than one-third of the local electorate, that called for an end to siting deliberations. However, doubts began to emerge concerning the credibility of the petitions, including the strong possibility that pressure tactics had been used to prod people to sign. In response, local authorities chose to keep open lines of communication with the crown corporation, and an April 1991 survey indicated that nearly three-fourths of local residents supported the facility.[14] All of this, facilitated by ongoing discussions with crown corporation authorities, led to the successful September 1991 referendum and the February 1992 decision to select Montcalm if environmental assessment reviews were favorable, which they were.[15]

Since the agreement was reached, comanagement has remained a central component in all facility construction and management plans that are being developed. Somewhat beyond the arrangements for public participation in Swan Hills, Manitoba has developed an across-the-board approach to ongoing public involvement in reviewing and shaping the direction of the facility. First, the Montcalm Council will appoint an independent Community Liaison Committee to "provide general oversight of the plant and to conduct any investigations with respect to the facility's environmental performance." [16] This goes well beyond the Swan Hills liaison committee, the membership of which is appointed by the crown corporation, not elected officials. Second, the Montcalm Council will nominate community representatives to a Plant Co-Management Committee that will also include senior plant management and will review and discuss ongoing plant activities. Third, the council will nominate two regular members to the board of directors of the Manitoba Hazardous Waste Management Corporation. Fourth, programs regarding protection of land and agricultural product values, environmental monitoring, employee training, and emergency response will be developed jointly by the corporation and the community. Fifth, Montcalm is "guaranteed participation in decisions related to any future expansion or significant change in technology associated with the facility that would result in an amendment to the operating license." [17] All expenses related to community involvement are to be covered by the crown corporation.

Institutional Change

The commitment to continuing public participation coincided with a pattern of institutional development that was also designed to foster trust. Manitoba followed the Alberta path in creating a crown corporation that would be set apart from provincial regulatory bodies. The Manitoba Hazardous Waste Management Corporation was created in 1986 and was responsible for establishing the voluntary siting process. Manitoba left open the possibility that its corporation would take full ownership of any facility that would be developed, contrary to the public-private partnership arrangement set forth in Alberta. In addition, the corporation would consider siting only on land that it could purchase, thereby enabling it to "assume, in perpetuity, the environmental stewardship of wastes and residues placed in the Corporation's custody." [18]

The corporation was quickly able to establish itself as a credible participant in hazardous waste policy debates, thereby insulating it from the considerable political turnover that was occurring in the province. Six years of New Democratic party governance came to an end in 1988 when the Progressive Conservatives were able to form a minority government. Two years later, the Conservatives won a narrow majority in a provincial election, headed by premier Gary Filmon, formerly a dean of a local business school. Despite these shifts and the ideological differences between these parties, the provincial approach to hazardous waste facility siting and management remained largely unchanged.[19] Policy stability may have been further reinforced by the Manitoba economy. Although mired in recession in more recent years, it has experienced few of the extreme cycles — and occasional downturns — of Alberta.[20]

As in Alberta, the crown corporation was intended to stand apart from environmental regulatory authorities, thereby precluding the perceived conflict of interest that can assist in scuttling regulatory approaches to siting. As one senior crown corporation official explained, "Operationally, the Corporation has established itself as a small project management organization. It functions independently of the developmental structure of government. This has been of major importance in establishing ourselves as a credible development proponent in dealing with our regulators, our industrial clients, the waste management industry and the general public." The corporation attempted to maintain much of the informal, small-scale focus of its Alberta counterpart, placing an emphasis on hiring staff with considerable professional experience and expertise in risk communication. In all of its written and verbal communications, the corporation insisted upon usage of clear, understandable language, so different from the highly technical, often obfuscatory language used in the cases involving the regulatory and market approaches to siting. The corporation also employed many of the key qualities that American public management analysts deem crucial for the successful "reinvention of government." [21] As one official who has worked for the corporation since its creation noted, "We were given a lot of autonomy from the rest of the provincial bureaucracy and all of its rules. We formed a small project team of twenty, people with diverse talents and skills. We were given lots of freedom to experiment, and freedom to fail, just like the literature on management excellence says is so important."

Crown corporation officials contend that no single format or blueprint was drawn up for developing the public participation mechanisms. Instead, they decided to experiment with different types of presentations and settings for meetings. Over time, they increasingly favored smaller open houses to larger public hearings. A variety of strategies were used to inform people of these opportunities and to encourage them to attend. But, above all, the corporation officials remained open to experimentation and avoided handing down any department or corporate party line, as is so common in regulatory and market approaches. "Often we would come home from a meeting at 1 a.m.," recalled one official. "We'd be exhausted but we'd recognize we needed to do something else. We would brainstorm and have the freedom to try something else. Having all this independence from elected officials and the environment ministry made all the difference. The only way they could get to us was through our board. And the board tried to give us as much room as possible without interfering."

As was the case in Alberta, officials who participated in the Manitoba process emphasized that it was crucial for the crown corporation — or any related siting entity — to devise and implement its own approaches to public participation. Whereas some promising American experiments have floundered in part because these duties were farmed out to a separate, private firm, Manitoba officials deemed it essential to keep these functions fully linked to the rest of the siting process. "It's so important to keep this in house, to keep the ideas and the information accurate and consistent," explained one official. "This sort of thing should never be turned over to some public relations firm, while the province [or state] does the rest of the siting process." The corporation staff was kept intentionally small, with about twenty employees; emphasized flexibility in all operations; and encouraged input from all staff.

Manitoba was able to take full advantage of the institutional structure because of the presence of capable policy professionals representing the corporation and local communities such as Montcalm. Crown corporation officials, particularly those on the front lines in the public participation process such as Alun Richards, won accolades similar to their Alberta counterparts. They also had to be effective in bilingual communication, as the Montcalm area includes many Franco-Manitobans, making it one of the larger French-speaking enclaves in central Canada. Furthermore, the reeve (tantamount to mayor) of Montcalm played a pivotal role in building public trust and standing as a credible broker between the crown corporation and the community. Montcalm Reeve Florent "Flo" Beaudette was an ideal figure for such a role. His family had farmed the same land for more than a century, and he was vice president of a local vegetable processing company. His standing took on added luster because he had served in Paris as a Manitoba trade emissary. All of this, in the words of one observer, made him "the perfect, perfect person" to foster deliberation over siting in an atmosphere of trust.[22]

A further advantage for the Manitoba crown corporation, particularly in comparison with its Alberta counterpart, was that it based its projected treatment and disposal needs on more reliable estimates of the volumes and types of wastes generated each year in the province. Whereas Alberta moved into the facility planning process in the absence of comparable projections, which has made it difficult for the facility to operate at full efficiency, Manitoba used its more exact data in deciding to scale back the scope of the planned facility. The decision by the crown corporation that inadequate waste existed to justify construction of an organic waste incinerator made the voluntary siting process somewhat easier, as it eliminated from consideration a waste management technology that many citizens and communities find particularly objectionable. This more careful planning may also assist the corporation with its long-term goal of seeking private investors for facility development and management and for other possible ventures in environmental management.

Compensation and Protection

Neither Montcalm nor the four other communities that actively explored the possibility of hosting Manitoba's comprehensive facility were in such serious economic straits as was Swan Hills. For Winnipeg, addition of the facility would have made little impact, given its diverse economy and population of 561,000. The main attraction of the facility to supportive residents was not economic but the prospect of safe, immediate access to sophisticated waste treatment and disposal capacity. "I'm surprised by how strong it [support] is among my fellow councillors," explained Winnipeg councilor Terry Duguid. "We're living with this stuff now and rather than dumping it in our sewers or landfill, it will be better to have a facility to dispose of it properly." [23] Incentives for the other communities were more mixed, as each saw some potential for economic diversification. But unlike traditional siting strategies, compensation was linked to other aspects of the siting and waste management process, not dangled as a singular enticement to overcome local opposition.

Perhaps the prospect of economic advantage from site acceptance was greatest in Hamiota, which had suffered significant declines in its predominant agricultural industry in prior years. "We've got to get industry," explained Mayor Don Mennie. "Agriculture will not sustain the area much longer. This area will not be here if nothing changes." [24] Such arguments were not, however, sufficient to convince a majority of Hamiota voters to support the facility in a 1990 referendum.

Montcalm was also eager to diversify its economic base, although it was not in as serious a situation as Hamiota. Its 1988 unemployment rate was 4.1 percent, well below the provincial average of 7.8 percent. Moreover, it ranked eighth out of twenty-one rural Manitoba census districts in per capita income in 1988. Thus it was hardly driven to cooperation by economic desperation. The three communities that comprise the Rural Municipality have a variety of services and businesses, and vegetable processing is a major activity. In the surrounding rural area, a mixture of crops are grown, including barley, canola, flaxseed, field peas, lentils, and sugar beets, in addition to the Manitoba staple of wheat. Nonetheless, Montcalm had experienced some economic and related population decline in prior decades and was receptive to the prospect of economic diversification made possible by facility hosting.

Unlike Swan Hills, Montcalm secured a portion of tax revenues from the facility, estimated to be between $145,000 and $160,000 a year during the construction phase, although dropping significantly thereafter.[25] Construction will lead to substantial short-term employment and spin-off activity, and facility operation will require about thirty to thirty-five full-time positions. The corporation estimates that about two-thirds of these positions will be filled by local residents. As in Swan Hills, the corporation has become a visible player in civic affairs, through measures such as contributing funds to help preserve a historic building slated for demolition, constructing a community center, and supporting local sport teams. Moreover, the Rural Municipality of Montcalm and Manitoba Hazardous Waste Management Corporation Co-Management Agreement requires the facility operator "to maximize opportunities for local residents and businesses to benefit from its hiring, purchasing and contracting requirements." [26] Further aspects of the compensation package remain to be fully negotiated in a separate agreement that is intended to guarantee nearby property owners, including farmers, of financial compensation should any facility operations "have a demonstrably negative impact on property values or crop prices." [27] As in Alberta, provincial and local officials are hopeful that the availability of secure, reliable waste management facilities will "be a major attraction for North American industries in their future location and expansion decisions." [28]

Environmental and public health protections, as in Alberta, are a related component of the siting process. Many provisions of the comanagement agreement involve measures to include the public to assure safe operation of the plant. An additional area that will require a separate agreement is the creation of an environmental monitoring program, the design and scope of which is to be established by the community and the corporation.[29] The corporation has also agreed to provide a used motor-oil and household hazardous waste drop-off station in Montcalm that citizens can use free of charge. When combined, these various provisions offer a variety of direct economic benefits, long-term economic assurances against loss, and commitment to safe facility management. As noted in the Alberta case, the mere promotion of compensation and protection alone is not likely to be sufficient to assure siting agreements but can contribute to the prospects for cooperation when integrated with other pivotal factors.

Sharing the Waste Management Burden

Manitoba realized from the Alberta experience that concentrating its entire burden for provincial waste management on a single community through a comprehensive facility was neither politically nor managerially desirable. As with the other components of its siting strategy, Manitoba borrowed heavily from its western neighbor in attempting to distribute the responsibility for waste management more broadly. It also provided specific assurances to any host community that it would not become an easy target for wastes generated outside the province. Much like Alberta, Manitoba developed a three-point approach for assuring a more equitable distribution of responsibility for waste management than is characteristic of market or regulatory strategies, which fail to link siting with other central aspects of waste management.

IMPORT CONTROL. The question of Manitoba's willingness to consider importation of wastes from other provinces and states to any new facility became a central issue in siting deliberations, as is common in many cases. "This was usually one of the first things we were asked about," explained a crown corporation official who was active in all phases of the public participation process. " 'Will you bring in wastes from other provinces or states?' People are very antsy about this, very uncomfortable with the idea of siting if it means wastes coming in from other places." The issue of possible waste importation was one of the few factors that upset Pinawa's otherwise solid support for hosting a facility. The import issue was raised in several public meetings, and the local

newspaper editorialized: "Make no mistake. Our province is being eyed jeal-
ously by producers of deadly waste throughout North America, and if we're
not careful, the night sky will be aglow with something other than aurora bore-
alis." [30] This commentary prompted Pinawa mayor Marvin Ryz to respond
with a public letter in which he not only defended the desirability of hosting
the facility but also emphasized that "at no time has there ever been any consideration given to wastes produced outside of Manitoba." [31]

The statement was something of an exaggeration as Manitoba officials have
not formally ruled out accepting some imported wastes. Given the decision to
avoid incineration and maintain a smaller facility than the one in Swan Hills,
the Manitoba facility will not be able to manage all of the hazardous wastes
that are generated within the province. As a result, Manitoba officials would
like to enter into formal agreements with other provinces (preferably Alberta)
or states that would allow for the transfer on an interprovincial or interstate
basis of limited quantities of wastes that then would receive treatment or disposal in the best possible technical setting. The Manitoba facility, for example,
will not be able to handle PCB wastes. Provincial officials would be willing,
under a structured agreement, to send those wastes elsewhere, with a linked
acceptance of other types of wastes for which Manitoba would have particular
treatment expertise.

Such transfers, however, are intended to be limited in scope. The co-
management agreement acknowledges a willingness to accept out-of-province
wastes but states that "priority will always be given to Manitoba's wastes." [32]
Such sentiments are repeated with frequency in interviews with all involved
parties. In addition, as one official explained, "We included, at the request of
the public, a clause in the [co-management] agreement whereby we could
publicly revisit the question of imports at any future point. That leaves open
the option of completely closing off the facility to out-of-province wastes
should there be need and support for such a step."

This form of import control does not go as far as Alberta's ban on imports
and exports but may offer a reasonable model for other provinces and states.
It provides far more assurance against exploitation from massive imports than
traditional siting agreements but leaves open flexible arrangements that allow
wastes to be moved to the most appropriate point for final management within
a geographic region. Some multistate agreements on hazardous waste management have been drawn up in the United States, but they are largely pro
forma arrangements designed to convince the Environmental Protection
Agency that capacity assurance goals have been reached so that the flow of
Superfund dollars can continue. Such an approach might also prove more

judicially acceptable than more recent unilateral state efforts to block all waste imports through various regulatory tactics, most of which have been declared unconstitutional because of the prevailing U.S. Supreme Court view of hazardous waste as an item of interstate commerce. And consistent with Manitoba's emphasis on public participation, it holds out the option of continued public consultation and possible termination of further imports, thereby adding assurance to the core provincial commitment to give priority to wastes generated within Manitoba boundaries.

REGIONAL STORAGE FACILITIES. While import restriction provides a guarantee against exploitation, the creation of regional facilities for waste storage will convince a host community that it will share the waste management burden with other communities. Manitoba has pursued this strategy with even greater vigor than Alberta. It has emphasized from the earliest stages of its voluntary siting process that construction of regional storage facilities would be integrated with the process of siting a large, central waste management facility. As in the Alberta case of Ryley, Manitoba officials emphasized that communities that seriously considered volunteering to host the central facility but were not ultimately selected would "be evaluated as potential sites for transfer stations." [33]

Well before the completion of the site selection process or construction of the Montcalm facility, the crown corporation put two storage facilities into operation. In 1988 the corporation took ownership from another governmental agency and substantially upgraded a facility in Gimli, a town of 2,000 residents located thirty-five kilometers north of Winnipeg, that can now serve as a storage facility for a large variety of hazardous wastes. These wastes may include PCBs, which will eventually be sent outside the province, as well as other wastes that will be shipped to Montcalm for treatment or disposal once that facility is completed. In 1989 the corporation opened a year-round hazardous waste collection depot in Winnipeg, one of the communities that actively explored the possibility of hosting the central facility. The depot is intended to store household and commercial wastes until they can be exchanged for other uses or eventually forwarded to Montcalm. During its first year of operation, in 1990, this facility collected more than thirty-seven tons of hazardous waste, more than some early estimates indicated was generated by all industry in the entire province.[34] Despite the substantial differences between Gimli and Winnipeg, the public has supported both transfer stations.

Manitoba intends to create additional storage capacity around the province, offering drop-off points convenient for the majority of provincial residents. It also has acquired transportation equipment suitable for moving waste about the province and holding household hazardous waste collection days in various communities, including rural areas, on a regular basis. These materials are collected free of charge from area households and are then stored at either Gimli or Winnipeg before reaching their final destination. Those materials deemed safe and reusable, such as used paint and painting equipment, are donated to social service organizations such as Habitat for Humanity.

The involvement of multiple communities has served to distribute broadly the responsibility for waste management in Manitoba. As a Winnipeg-based planning consultant noted, the hazardous waste issue "requires total community action and a recognition that hazardous waste must be considered in a healthy community context with collective responsibility for its disposal." [35] A major theme throughout the Manitoba siting process and in more recent stages has been the widespread nature of hazardous waste generation and the need for collective responsibility for proper waste management. The involvement of regional storage facilities, mobile collection units, and individual households and consumers removes some of the burden for waste management traditionally concentrated on the community hosting major treatment or disposal facilities.

WASTE RECYCLING AND REDUCTION. Provincial efforts to promote the recycling and reduction of waste help minimize the burdens imposed on Montcalm for waste management. Most provinces and states — like their respective federal governments — are supportive of waste recycling and reduction. But in practice few pursue these functions with much rigor or even begin to attempt to integrate them as fully into a comprehensive waste management system as has Manitoba. As with the regional storage facilities, serious pursuit of recycling and reduction initiatives was developed in advance of the larger siting effort. According to policy analyst Michael Heiman, Manitoba's "emphasis on optimum source reduction prior to any planning for facility location is significant." He noted that such efforts are essential to building broader political support for facility siting and for helping to assure that Manitoba "is not saddled with costly overcapacity in the face of variable future demand." [36]

Manitoba pursued its commitment to the "4Rs" of reduction, reuse, recycling, and recovery through a number of mechanisms in the late 1980s, several of which have been borrowed by other provinces and states in more recent years. First, the crown corporation established a generator services program to encourage waste reduction. This program provides free technical advice to corporations on a confidential basis and publishes an index of individuals and organizations with relevant technical expertise that can be hired. The program fields about one hundred inquiries a month, working with a wide variety of Manitoba waste generators. Second, the corporation developed the on-site program, which uses resources from a federal government unemployment program to place qualified unemployed individuals with private firms that wish to pursue waste reduction alternatives. Third, the corporation has developed a system for waste audits, whereby corporation officials inspect waste generation facilities. Instead of focusing on technical compliance with extant regulations, these visits are intended to explore options that might lead to significant waste reduction. Fourth, the corporation established a Manitoba Waste Exchange, a nonprofit information center intended to help industries find viable markets for wastes that have reuse potential. Fifth, the corporation devised a series of educational and informational materials that explained waste reduction and recycling options for large industry, small business, and consumers alike. All of this was presented in unusually clear, jargon-free language, including prominent listings of toll-free numbers that citizens could call to receive additional information.

Hundreds of Manitoba firms and communities have taken advantage of these services, often with significant waste reduction results. For example, a consumer packaging company located in Winnipeg has experienced dramatic reduction in the volumes and types of hazardous and solid wastes generated, in part, through participation in these programs. Gravure Graphics employs about one hundred people and makes food and consumer packaging for wide distribution in Canada and the United States, using materials such as polyethylene, polypropylene, and cellophane. Package redesign has led to significant reduction in the amounts of these materials being used. In turn, the company has dramatically reduced the amounts of solvents and inks that it produces. These constitute the main hazardous wastes generated by the firm; in 1989 alone it created more than sixty barrels of liquid hazardous waste a month, much of which had to be exported outside of Manitoba for treatment. But through a variety of waste reduction efforts, many under the auspices of provincial programs, this number dropped to forty barrels a month in 1990 and to five to twelve barrels a month the following year. Most of the measures used to promote this reduction were highly nontechnical, and many have resulted in considerable cost savings for Gravure Graphics. They include purchasing only the quantity of ink required to complete a job and saving leftover ink for another run or job using similar ink; using solvent-dampened rags and cloths instead of substantial amounts of solvent; training workers in alternative procedures; examining areas of the plant that produce waste solvent to determine how it might be reduced; using nonhazardous products for cleaning purposes wherever possible; and promoting an aggressive employee recycling program on the site, with any proceeds directed to the employees' social committee fund.[37]

Manitoba's waste recycling and reduction programs were operational and well known in the province before the final decisions on facility siting were made. The corporation in 1988 completed a thorough analysis of the volumes and types of wastes generated in Manitoba. When combined with the waste reduction and related initiatives, it gave provincial officials a much clearer sense of waste treatment and disposal capacity needs than in most provinces and states, which provided essential reassurance to Montcalm and other siting finalists that they would neither be exploited nor become the only community in Manitoba — or central Canada — expected to assume an active role in hazardous waste management. Manitoba thus demonstrated that the historic siting agreement in Alberta was not an accidental stroke of luck. ·

North Carolina: The Greensboro Exception amid Nimby Chaos

North Carolina would, in many respects, seem to be the least likely place in either the United States or Canada to offer a model for siting, with many similarities to the breakthrough agreements in Alberta and Manitoba. Both regulatory and market siting approaches in the state have failed dismally. Among the more recent of a long series of setbacks was the rejection — by 93 percent of participating Pender County residents — of a September 1991 referendum proposing a hazardous waste incinerator. Not surprisingly, North Carolina continues to export the majority of wastes that cannot be disposed of on-site, reaching 81 percent of its total in 1990. Environmental policy analyst William Gruber characterizes the state's approach to hazardous waste as a "we make it, you take it" policy.[38] Others use stronger language, particularly officials and citizens of states that ultimately must assume responsibility for large quantities of North Carolina hazardous wastes. According to the press secretary for South Carolina governor Carroll A. Campbell, Jr., "We remain in pitched battle over North Carolina's intransigence and unwillingness to honor its own commitments to site facilities." [39]

Amidst all these problems of failed siting efforts and growing tension over the waste export issue, a singular siting success in North Carolina from the mid-1980s shatters the conventional understanding that siting is an impossibility in the state. The resulting storage, treatment, and disposal facility, located in Greensboro, is sufficiently small to handle only a portion of the state's off-site waste but serves as a reminder that siting need not be an impossibility even in a contentious setting. And given its strong parallels to the Alberta and Manitoba agreements, the case further suggests promise for the voluntary siting approach.

The Greensboro case differed from those in Alberta and Manitoba both in size of facility and size of volunteer host community, given its population of 184,000. The case was also distinct because of its comparatively reduced role for governmental authorities. No statewide public participation process was pursued; no crown corporation was established to facilitate construction. Nonetheless, the North Carolina case involved a voluntary siting effort, in which a single entrepreneur candidly and widely shared his proposal with community leaders and made clear he would proceed only if he could earn their support. After prolonged negotiation involving many revisions of plans by the developer, the facility was approved and received a federal RCRA permit to begin operations in July 1985.[40] Located in a heavily industrial area, the ECOFLO facility offers a combination of treatment and disposal methods, including blending, bulking, centrifugation, decanting, filtration, neutralization, oxidation and reduction, precipitation, and solidification and stabilization. However, the plant does not incinerate hazardous wastes, and while it accepts a fairly wide variety of wastes, it does not accept PCBs, dioxins, or cyanide.[41]

As in the other successful cases, early and extensive public participation of Greensboro area residents was essential to the cooperation that emerged. Unlike other North Carolina siting cases, the ECOFLO proposal was not handed down from above as a fait accompli. Instead, Tom Barbee, who had business experience in Greensboro as a waste transporter, attempted to interact with a wide range of interested individuals and groups about his proposal. In these interactions, he made clear that he was not adamant about imposing his preferences for any facility on the community and that he wanted its final design — and any final decision — to emerge through broad community consensus. According to environmental policy analyst Frances M. Lynn, "Barbee talked and touched as many bases as he could (ministers, neighbors, the planning and zoning department, the county commissioners)." In addition to his numerous meetings with individuals and organizations, added Lynn, "Barbee sponsored his own public meeting — the first of its kind — which preceded the state's public hearing." [42]

Much like the Swan Hills and Montcalm cases, such frequent and open access to the site proponent was essential to the development of trust. Lacking the larger infrastructure of the provincewide public participation efforts, Barbee had to, in essence, develop his own system in a state where animosity toward hazardous waste siting proponents was widely known. "There were no magic moments," claimed Barbee. "There was just ongoing communication . . . [including] informing the public . . . and asking them to be involved in the planning process." [43] As a result, Barbee forged ahead, consulting widely and searching for the sort of facility design that could win public support and also provide a needed waste management service.

Two aspects of the participation process appear to have been essential. First, Barbee emulated the pattern of other successful policy professionals, reversing the typical, top-down pattern that pervades regulatory and market siting efforts. According to Lynn, one Greensboro official characterized Barbee as a "dream" because of his style of operation; a fire official was equally impressed, seeing him as different from traditional hazardous waste managers or facility proponents who "come in as if they are in a foreign country." Barbee, a Greensboro native, had the advantages of knowing his home community and having worked in waste transport in the town. But he also found a way to approach the subject of facility siting — including risk communication — in a way that could win public trust, foster prolonged dialogue, and result in a siting agreement. In a 1984 local television interview, he came off as entirely credible when he said, "I'm a resident of this city too and I plan to stay here," adding that his company "honestly wants to be a service to the community. . . . We want to help local communities handle their waste as responsibly as possible. . . . We are on the side of environmentalists." [44] Most waste management officials — in either the public or private sector — would meet extreme skepticism from the public upon making such a statement. Barbee was believable because his actions and commitments to an open planning process matched his rhetoric, as was the case with his counterparts in Alberta and Manitoba.

Second, Barbee's efforts received an important institutional boost from the Guilford County Hazardous Waste Task Force, which was initially formed in 1979 through a small grant from the EPA Office of Public Participation. [45] Although overshadowed by programs such as Superfund and RCRA, this modestly funded effort was intended to advance "public understanding of hazardous substances and proper waste management through educational efforts, to review laws and their implementation, and to make suggestions or to assist officials where this seemed to be appropriate." [46] The task force took this charge seriously, continuing to operate with state and local resources after the Reagan administration terminated funding for the program in the early 1980s.

The task force played an active role in community education on hazardous waste issues, offering public meetings, informational displays at shopping centers, and short courses for public school science teachers and other individuals.[47]

But the role of the task force went beyond pure provision of information to the public. When Chemical Waste Management advanced, in secretive fashion, its proposal for a major waste disposal facility for Greensboro in 1983, the task force assembled a "reading committee" to review its proposal. This committee consisted of a wide range of citizens, including a number with considerable expertise on matters related to hazardous waste. They raised a number of important objections to the proposals that contributed to the state's decision to reject the project.[48]

In the ECOFLO case, the task force once again formed a reading committee but soon realized that it was working with a very different type of facility proposal and proponent. Task force members reviewed the proposal and discussed it with area residents, ultimately making suggested changes in facility design and operational procedures before any permit application would be sent to the state. This process was facilitated in large part by Barbee's unusual willingness to share information with the task force and the public and respond seriously to their proposals. The work of the task force was also supported through unusually strong cooperation from relevant Guilford County agencies, particularly the Emergency Management Assistance Agency.[49] The task force gave Barbee a visible, credible outlet to supplement his own public participation. Having been in operation for more than five years and familiar to many citizens for its prior education efforts and opposition to the Chemical Waste Management proposal, the task force was an important player in this siting agreement.[50]

The public participation efforts and institutional innovations were supported by other factors that contributed to public support for the siting agreement. Economic compensation was not a significant factor, as ECOFLO would neither employ nearly as many people nor offer the sorts of spin-off economic benefits as its counterparts in Swan Hills and Montcalm. But explicit guarantees against exploitation proved to be central to the resulting public support that the ECOFLO proposal received.

These protections took two distinct forms. First, Barbee made numerous adjustments to his initial proposal, many of which enhanced the prospects for safe management of any wastes that would enter the facility. These involved alteration of planned transportation routes to the facility, installation of an extra fire protection wall, training of a special fire brigade in case of an emergency, and a mutual aid agreement with other area firms should an emergency occur. In addition, ECOFLO agreed to refrain from using certain technologies or accepting certain materials thought to pose too high a public health risk.

Second, Barbee emphasized that the facility was intended to serve primarily local and intrastate waste generators. He stressed that he wanted to be a primary point for waste management for Greensboro area waste generators, many of whom were having difficulty finding places to send their wastes. He characterized his clients as "little guys who need environmental expertise but can't afford to have their own full blown staff and don't produce enough waste to get the big handlers interested in them." [51] Under federal law, Barbee could not guarantee local residents that he would serve only local clients, but he was successful in arguing that any ECOFLO facility that might open in Greensboro would not become a magnet for massive quantities of wastes from other states.

These factors combined to give ECOFLO a strong base of support and has allowed it to begin to make a significant contribution to the waste management problems in central North Carolina. While the Greensboro case lacked the large, structural features necessary to develop the voluntary process and build support for a major facility, such as those in the two Canadian provinces, many of the same lessons for siting success can be drawn. These include the commitment to early and extensive public participation, institutional reform to foster public dialogue and trust, and special commitments to protect the public — both through added safety provisions and a commitment to focus on a limited volume and series of wastes that were primarily to be of local origin.

Minnesota: Voluntarism, Nimby, and a New Commitment to Prevention

Minnesota has a history of hazardous waste management and facility siting that bears striking resemblance to Alberta's. The state generates about the same amount of hazardous waste per year (approximately 225,000 tons) and has had a tradition of failure in its efforts to site commercial hazardous waste treatment and disposal facilities. Historically, many Minnesota wastes were dumped in a variety of slipshod ways, as in Alberta, although in more recent years the state has tightened regulation and relied on a combination of on-site storage and disposal and out-of-state export.[52] State officials were eager to break through a pattern of Nimby opposition in the late 1970s, just as their counterparts were in Alberta.

Minnesota attempted two variations of a regulatory siting strategy in the late 1970s and early 1980s, with predictable results. In 1975 state officials selected four candidate sites without prior consultation but were forced to withdraw them from further consideration in the face of vehement public opposition. Five years later, under a new Waste Management Act and Waste Management Board, the state used new technical criteria to screen the entire state and again come up with a series of possible siting candidates.[53] But another bout of strong public opposition led the state to suspend its siting process in April 1984 and reinvent its siting strategy once again.

Desperate for a better way to conduct siting, and increasingly concerned about the continued availability of export as an option, Minnesota officials amended the Waste Management Act in 1986 with a so-called voluntary approach that was, to some degree, modeled on the Swan Hills experience after consultation with Alberta officials. As in Swan Hills, siting would only be done on a voluntary basis and any host community would have considerable input into facility design. Generous compensation packages would be part of the negotiation process, ranging from economic support for local governments to added safety provisions. Furthermore, the state agreed to explore the possibility of playing a direct role in construction and operation of any new facility, to ensure residents against the risk of exploitation from a private firm that might walk away from a facility if problems arose.

Many of these features made success more possible than anything else Minnesota had tried. Multiple volunteer counties did come forward and explore the possibility of hosting. Prolonged dialogue occurred between state officials and many community leaders. Two in particular, Koochiching and Red Lake Counties, both in the northern part of the state, became serious candidates in 1988 and 1989. Eventually, however, the siting process collapsed. Koochiching withdrew in March 1989, facing a divided county board and considerable citizen opposition.[54] Red Lake County voters rejected a referendum on a proposed facility by a 69-to-31 margin in November 1990. Strong, locally based political opposition emerged in both counties, as larger environmental groups that attempted to participate in the opposition were actively discouraged from doing so by local residents. "Both are face-to-face communities who don't like to be told what to do from outside, professional groups," recalled one state official. "At one crucial meeting in Red Lake, a hockey stick was placed between door handles to bar access to anyone who wasn't a resident. They wanted to make their own decisions." Neither Koochiching nor Red Lake chose to continue in the siting process.

Although modeled somewhat on the Alberta experiment, the Minnesota experience does not represent a full test of the voluntary approach to waste management. Several important distinctions between the Minnesota approach and those developed by both Alberta and Manitoba may have undermined the possibility of a siting agreement that appeared to exist in the earlier stages of the process.

First, Minnesota's approach was voluntary only in the sense that it agreed not to force any community to accept a facility. No massive public information and education effort was conducted before seeking volunteers. Unlike Alberta and Manitoba, the state began its efforts by extending an open invitation to all Minnesota counties to explore the possibility of hosting a site. It placed particular emphasis on the prospect of financial support for participants. This offer of support took the form of a $4,000 monthly stipend, which could be expanded to $6,000 a month as the process proceeded. The money was to be used to cover any local costs related to participating in the siting review process, such as forming citizen advisory and liaison groups. To receive the funds, a county board just had to pass a nonbinding "resolution of interest" in possible facility siting.[55] Only at this point did the public participation efforts of the Waste Management Board begin, including the sorts of public meetings common in Alberta and Manitoba. In addition, some local citizen delegations, such as a group from Red Lake County, were taken to Swan Hills to see the Alberta facility in operation and to Blainville, Quebec, to see its newly opened facility.

Thirteen counties passed resolutions of interest and began to meet with state officials. However, most quickly withdrew from further consideration, leaving four counties that reached the stage of test drilling and soil sampling to determine the technical feasibility of siting. Two of these counties, Nobles in the south and Stearns in the north, dropped out in 1988, leaving only Koochiching and Red Lake Counties.[56] In both settings, some visible, outspoken support existed for continued exploration of siting, but a pervasive set of doubts about the process and viability of the proposed facility led to opposition. Unlike the successful sitings in the Canadian provinces, no strong base of trust developed through early and extensive participation and dialogue could be drawn upon at key stages of the process. "They [the Minnesota siting officials] didn't do nearly as much as we did in terms of education, of getting people really involved in the process early on," explained a Swan Hills elected official, who consulted with Minnesota officials, met with delegations of Minnesotans who visited the Swan Hills facility, and was invited to speak before possible host communities in the state. "People there felt like it was being forced on them because they hadn't been brought in the way our people were."

Second, Minnesota encountered huge problems in developing a public agency that could capably and credibly handle all aspects of the siting process, including risk communication. Just as effective institutional reform was crucial in Alberta and Manitoba, its failure in Minnesota undermined any prospects for an agreement. The state Waste Management Board was responsible for all aspects of the state siting effort but experienced scandal and considerable staff turnover at key points of the negotiation process with interested communities. In April 1988 board chair Joseph Pavelich unexpectedly fired deputy chair Tom Johnson and staff member Susan Thornton, both veteran employees who had been active in the development of the siting process and its early stages of implementation. This triggered great controversy in the local media, leading to a resolution by the Minnesota Senate Environment and Natural Resources Committee for the full board to review the firings.[57] The board did so, urging Pavelich to rehire the employees; he refused. By October the board was dismantled, reorganized, and placed under the larger auspices of the Office of Waste Management within the Minnesota Pollution Control Agency. Pavelich was transferred to another prominent state government job, leading to still additional staff changes. Finally, an April 1989 report from the legislative auditor revealed irregular spending practices by the board before its dismantling, including excessive overtime payments, unnecessary travel expenses, and sloppy management practices.[58] Repercussions from these and the other controversies led to an overall staff reduction from eleven to four full-time employees, leaving Minnesota far fewer and less experienced policy professionals than in Alberta or Manitoba. Further controversy erupted when news was revealed that several of the firms under serious consideration for facility construction had extremely suspect environmental safety records in some of their other operations.

All of these developments occurred in the midst of state efforts to prove itself worthy of a county's trust to site and construct a major hazardous waste disposal facility directly or through private contractor. Whereas Alberta dealt with personnel controversies with relative aplomb, any prospect for Minnesota to reach an agreement with a county was undermined by the dismal performance of the agency—and key policy professionals—responsible for the process.

Third, Minnesota officials offered varying explanations throughout the process of the extent to which any facility would accept wastes from outside of Minnesota. Initial discussions emphasized that controlling imports would be an impossibility, based on constitutional reasons and Minnesota's desire to keep open existing export options. Later discussions, which emphasized the state's growing interest in building its own facility, assumed that as a public entrant into the market it could block or restrict imports and was prepared to do so. As one prominent state official wrote in mid-1988, "Citizens have expressed repeated concerns that the facility would receive waste from across the country. Citizens feel a responsibility to care for wastes generated in-state, but a lesser duty towards wastes from other regions." [59] In response, a strategy involving public facility ownership and a contract to a private firm for day-to-day facility operation was given serious consideration, as it might bolster the legal case to control imports as well as to provide additional safety reassurances to the public.[60] The Waste Management Board report noted, "Successful siting of the facility is partially dependent on assurances to potential site areas that the facility will not be 'national' in scope. . . . State ownership provides greater opportunities for such assurance." [61]

But in the final proposal made to Red Lake County, the import option was left open. Voters were asked to consider construction of a waste treatment facility that would accept up to 24,000 tons of inorganic hazardous waste a year. Through previous negotiation, the state had dropped its earlier proposal to include an incinerator as part of the facility. A generous, multifaceted compensation package was included, with the most unique feature a "property value assurance program," designed to assure area property owners that they would receive fair market value for their property if, for any reason, the facility triggered a decline in property values. But on the import issue, the agreement made clear that the facility would "accept waste produced in the state of Minnesota and waste produced in other states." It did note that "priority" would be given to Minnesota waste but, unlike Manitoba, made no permanent commitment to such prioritization.[62] The agreement also left no provision for the public to reopen the issue if waste imports became a later concern, which was central to the Manitoba siting agreement. Furthermore, no efforts were made to formally link the siting effort with other burden-sharing arrangements, such as regional transfer stations.

These important distinctions between the Minnesota case and its counterparts in Alberta and Manitoba suggest that a fully voluntary, burden-sharing approach has yet to be attempted in the United States. However, the Minnesota case probably comes closest, with the possible exception of Greensboro, among American siting experiments. And, for all of its resulting administrative problems and Nimby outcomes, Minnesota experienced some degree of success in siting once it abandoned its two attempts through a traditional regulatory approach. Multiple volunteer communities did come forward and at least two gave serious, prolonged consideration to the possibility of hosting the facility. Given all the scandals surrounding the Waste Management Board and related complications, the Minnesota case hardly constitutes a full test of the replicability of the voluntary approach in the United States. But despite its problems, it came a good deal closer to building meaningful public dialogue and approaching a solution than the states and provinces that used the regulatory and market approaches. And, as one senior official noted, "We've learned a lot through this process. We made mistakes, but I think we'd have a good notion of how to do this again and be more successful."

From the ashes of the failed effort to site a waste management facility, however, has arisen an innovative alternative: waste reduction as the state's primary waste management policy. Minnesota remains highly reliant on exports for waste management that cannot be completed on-site. While the state retains the option of reactivating its existing siting process or designing an entirely new method, state officials seized the setbacks in Koochiching and Red Lake Counties to make an all-out effort on waste reduction the central component of the state's hazardous waste management strategies for at least the next several years. This clearly reflects the general orientation of the Office of Waste Management, which has fully absorbed the duties held previously by the Waste Management Board.

Signs of the shift were evident before the decisive Red Lake County referendum rejection. One week before the vote, when Office of Waste Management officials were fairly sanguine about prospects for passage, a senior official pondered, "Maybe we should be taking all that money we're spending on siting and use it to work much harder on waste reduction. Minnesota doesn't generate all that much waste and most generators still have a lot of out-of-state options. Maybe reduction should be a bigger priority than siting." On the same day, another official, equally hopeful of a siting agreement, confirmed that "the more we begin to see that there's no simple place to put this stuff, the more attractive pollution prevention becomes."

The referendum vote drew added attention to a series of state hearings that were under way on methods to reduce regulatory, economic, educational, and institutional barriers to pollution prevention. Office of Waste Management staff attempted to draw diverse constituencies, including industry, varied state agencies, and environmental groups, to the hearings. They released a series of reports in 1991 that were based on the hearings and made a strong case for shifting from siting to prevention as the state's main waste management approach.[63]

The emphasis on waste reduction was accompanied by a set of new measures that transcended those possessed by other state and provincial governments.[64] Previously, Minnesota's lone program with a preventive focus was the Minnesota Technical Assistance Program (MnTAP), which has provided technical assistance in pollution prevention to industries located within the state since the mid-1980s.[65] MnTAP was expanded under the new approach but also significantly supplemented with other programs, including the Minnesota Toxic Pollution Prevention Act.

This legislation specifies the prevalence of cross-media (air, water, land) transfers from pollution control and waste management efforts as a central reason for pursuing pollution prevention with unprecedented vigor.[66] All Minnesota facilities required to report chemical releases under federal right-to-know requirements (incorporated into the Superfund program) must submit "toxic pollution prevention plans" for review by the Office of Waste Management. More than 500 Minnesota firms must comply with these requirements, which call for preparation of comprehensive plans that outline their current use and release of toxic pollutants. Many of these pollutants are classified as hazardous wastes under RCRA. The firms must also establish formal goals for eliminating or reducing their use and release of these substances. The plans must be updated on an annual basis by reports that describe and certify progress made toward meeting each pollution prevention performance goal specified.[67] Minnesota officials are developing supplemental state requirements that may require eligible firms to either release additional data or set more exact timetables for reducing toxic releases. "These plans present a whole new wrinkle for us in our pollution prevention efforts," explained one state official.

The Minnesota program has also developed a novel mechanism for funding future pollution prevention efforts through fees on facilities that release toxic substances. The fees, which are expected to be increased over time, are designed to provide fiscal incentives for waste generators to reduce waste. They are also intended to provide considerable funding for state pollution prevention programs, thereby reducing prior reliance on general state revenues.

Other central components of the state's waste reduction strategy include integrating traditional medium-based permits and inspection to minimize cross-media transfer and promote waste reduction wherever possible. This project began on a pilot basis in 1993 involving thirteen large industrial plants along the Minnesota shore of Lake Superior. Unlike the fragmented process that focuses on technical compliance with medium-based standards, the Minnesota effort gathers teams of inspectors from various specializations to examine jointly all aspects of facility operation. The teams are expected to promote pollution prevention, according to one official, "in all interactions with these facilities" and offer technical assistance on any pollution prevention options that might emerge. The state is also implementing a flexible permit program, which offers regulated firms greater compliance flexibility in exchange for considerable overall emission reductions and monitoring system improvements. These initial efforts are expected to be expanded, thus making pollution prevention an integral part of the statewide regulatory process.[68]

Furthermore, state officials have agreed to complete a thorough review of their own environmental performance, seeking ways to reduce their generation of hazardous waste. A 1991 executive order requires all state departments and agencies to draft and implement pollution prevention plans, which must provide extensive information on the volumes and types of toxic substances released into various media and ways to reduce them. The process is overseen by the Office of Waste Management and has already resulted in some significant discoveries of waste reduction methods.

Finally, Minnesota has built upon the MnTAP approach with an array of initiatives designed to raise awareness of and build support for pollution prevention. These include regular workshops and public meetings, an annual statewide conference on pollution prevention, Governor's Awards for Excellence in Pollution Prevention that are granted to three Minnesota firms a year, household hazardous waste programs, small matching grants for firms eager to experiment with pollution prevention approaches, and curricular packages for use in public schools. These efforts have essentially shoved the issue of siting to the side, while a variety of state agencies — environmental and other — and firms have attempted to demonstrate their commitment to the new waste reduction effort. Even Governor Arne Carlson has become an outspoken campaigner for this approach, proclaiming in 1991 that "the future in Minnesota is in preventing pollution at its source."

Pollution control efforts were embraced and expanded in 1993 legislation that amended the Minnesota Waste Management Act. Key provisions included increasing the number of firms required to pay fees and complete pollution prevention plans, labeling requirements for numerous hazardous products, and banning the continued disposal of many common items, ranging from auto antifreeze to fluorescent lamps, as solid waste. The legislation also broadens the definition of materials deemed as hazardous.

Minnesota officials reported a 35 percent drop in state-generated hazard-ous waste between 1988 and 1992, with additional declines expected in future years. Some individual firms have achieved particularly dramatic reductions. Conklin Company, a manufacturer of agricultural, cleaning, coatings, and lubrication products, reduced its generation of xylene-based hazardous wastes by 13,653 pounds, or nearly 70 percent, in a single year. The Bureau of En-graving, a manufacturer of printed circuit boards, eliminated methylene chlo-ride and 1,1,1-trichloroethand from its processes, more than 650,000 pounds a year.[69] However, most officials assume that considerable, albeit significantly lessened, volumes will continue to be generated and that at some point the state will have to revisit the siting issue. This will be especially true as Minne-sota's expanded definition of what constitutes hazardous waste leads to volume increases and if, for any reason, Minnesota loses access to its waste export sources.

Some officials remain particularly interested in the Alberta and Manitoba cases, eager to use these experiences to refine the flawed voluntary process that Minnesota used. The officials contend that, despite the setbacks of the 1980s and 1990, a workable siting process could be developed, particularly after the state has made a full effort to minimize the volumes of waste being generated. But, in the meantime, the concerted shift from a siting to a reduc-tion emphasis offers parallels with the successful siting cases, including wide-spread public participation and broad distribution of the responsibility for waste management.

California: Promise for Burden Sharing Undermined by Bureaucratic Objection

California differs markedly from Minnesota in the volumes of hazardous waste generated and the number of abandoned waste sites that warrant serious attention. The nation's most populous state consistently ranks as a national leader in both categories. But much like Minnesota, California devised an innovative policy for siting and waste management in the mid-1980s that held considerable promise for development of a more equitable, open approach. That process has had at least some success in promoting greater waste reduc-tion and initiating some serious dialogue over equitable burden sharing of waste management. Overall, however, its considerable potential was under-mined by a 1987 California Department of Health Services (DHS) ruling.

Thus the California case presents a mixed record, leaving many unanswered questions concerning long-term plans for hazardous waste management.

California began to explore siting alternatives in the early 1980s, as the severity of its hazardous waste problem became more evident. Older facilities, including landfills, began to close, no new facilities had come on line since 1972, and recent market-style efforts to attain siting agreements failed.[70] Proposals began to surface in the state legislature to adopt a regulatory approach to siting, giving regional authorities the power to override local opposition to proposed sites. These were ultimately rejected in favor of legislation that sought a more balanced approach and placed a strong emphasis on equitable distribution of responsibility across the state.

Enacted in 1986, the Tanner act, named after Assembly sponsor Sally Tanner, established a comprehensive planning process that would precede specific siting attempts. Each California county (many of which are massive in physical size and have populations that exceed a number of individual states) was to prepare a plan that reported the volumes and types of hazardous waste generated within its boundaries and offered a proposal to assure safe management of those wastes. Counties were encouraged to devise partnerships to share the burden of waste management. An emphasis was placed on a "fair share" distribution of responsibility.[71] The planning process called for extensive public involvement, through creation of county advisory committees and local advisory committees, and was supported by $10 million in public funds. As environmental policy analyst David L. Morell noted, "This legislation provided incentives for the state as well as for localities to acknowledge and take responsibility for management of the hazardous wastes they were generating in their own areas." [72]

The planning process moved smoothly, with all fifty-eight counties voluntarily agreeing to prepare draft county hazardous waste management plans and most meeting deadlines for submission. The plans provided an unprecedented look into county-by-county waste generation and management. Much like the release of community right-to-know information concerning toxic substances, this process drew extensive public attention to the issues of local generation of hazardous waste, which created considerable impetus for generators to pursue waste reduction more aggressively than ever before.[73]

The process also encouraged counties to begin to take seriously their role in hazardous waste management. According to Morell, they "were excited about the prospect of having a plan to handle their hazardous wastes which eliminated the fear of preemption and disproportionate burdens. They were to rely on [the legislation's] 'fair share' provisions to protect them from having to accept unwillingly facilities which exceeded defined local needs." [74] At this point, California was forming a reputation for developing a model worthy of possible emulation by other states, just as Alberta would become the standard for later Canadian siting efforts.

Its stature as a national leader was short lived, however. The entire planning process began to unravel after the California Department of Health Services issued interpretations of the Tanner act that gutted the fair share provisions. DHS guidelines called upon all counties to refine their planning processes to develop siting criteria for all types and sizes of waste management facilities, independent of county waste management needs. This negated the central features of equitable distribution and the potential for intercounty partnerships. Instead, according to the binding DHS interpretation, all communities were to prepare to site a wide range of possible facilities, including massive ones that would serve state or regionwide needs.

DHS gave the California Appeals Board the authority to override local rejection of a proposed facility, allowing, in effect, the state to use the Tanner act as a "preemption tool." [75] The board was created under the legislation to provide appeal of rejected facilities that were consistent with approved county waste management plans; thus the legislation was not purely voluntary from the outset. Although the appeals board had not triggered opposition initially, its expanded role, in tandem with the elimination of fair share principles, effectively destroyed public trust in the process.

These steps gave DHS the more centralized oversight of the state waste management process that it sought but little effective capacity to initiate public dialogue and systematically explore waste management options. The more decentralized strategy envisioned under the legislation was particularly desirable given the economic diversity of the state as well as its huge population, which exceeds the entire Canadian population by about 3 million residents. Counties have continued to pursue planning efforts, in large part out of their sense that plan approval "would lessen their vulnerability to unwanted facilities from the appeals process." [76] However, state funding for these activities has been slashed and only twenty of the fifty-eight counties have had their plans approved by the state.

Moreover, subsequent efforts to site facilities have operated independently of county planning efforts and follow a familiar market-style pattern that culminates in a Nimby reaction. No new facilities have been opened, and conflict has prevailed at each new proposed site. California and its waste generators have continued to make significant strides in waste reduction, but the state has come under increasing criticism for growing reliance on waste exports.[77] In 1989, for example, California hazardous waste was exported to at least thirty-five other states and a number of foreign nations.[78] Despite the enormous unreliability and variation in calculation of waste movements, California has by all accounts one of the largest net waste export levels of all the American states.

The greatest unanswered and unanswerable question from the California case is whether the burden-sharing arrangement envisioned under the Tanner act would have been successful if implemented as designed. The coercive nature of the appeals board, even in the initial design, distinguishes it from purely voluntary approaches. Moreover, the fair share arrangements might have encountered constitutional problems, given the American penchant to view hazardous waste as a marketable commodity worthy of unrestricted movement across jurisdictional boundaries. Nonetheless, the early stages of the Tanner act implementation reemphasize the importance of equitable burden sharing in waste management. Under this principle, communities took their planning processes seriously and began to look carefully at waste reduction and other waste management options. In many respects, the brief period before the introduction of the DHS guidelines marked the highlight in California hazardous waste management of the past two decades. Some observers remain hopeful that this experience can lead toward a revitalization using similar principles developed in the 1986 legislation.[79]

Quebec: Siting Success, Implementation Debacle

Aside from the Alberta and Manitoba agreements, only one additional major waste treatment or disposal facility was approved and opened in Canada during the 1980s. Neither the siting process nor the facility ultimately constructed in Quebec was as comprehensive as Alberta's or the one being completed in Manitoba. However, the siting process leading up to the Quebec agreement does bear some resemblance to the approach set forth in the other cases. The province's failure to live up to its commitments has served to undermine support for the facility or a more comprehensive waste management system for the province.[80] Ironically, after all the turmoil surrounding the facility in recent years, provincial officials are giving serious thought to creation of a crown corporation and a waste management approach similar to that already developed in Alberta and Manitoba.

The Stablex Canada facility in Blainville, Quebec, was endorsed by local residents in 1981 and opened for business in early 1983. It was designed to serve as a dominant agent in provincial waste management for Quebec. The province decided to pursue a siting agreement in the late 1970s, moving beyond its earlier, purely market, approach to siting. The decision was prompted by revelations that Quebec waste disposal procedures were haphazard and that major water bodies had been seriously contaminated by wastes as well as other forms of pollution.[81]

Provincial officials selected Stablex to construct and operate a facility for inorganic waste, employing the Sealosafe technology that the firm uses in numerous sites in Western Europe. Inorganic hazardous liquids, sludges, and solid wastes are converted into a solid, stable product that resembles concrete and is placed in a landfill. The Stablex facility in Blainville lacks the capacity of the facility at Swan Hills to incinerate wastes, conduct a variety of treatment functions, or handle organic wastes such as PCBs, but it has the capacity to solidify and landfill more than 70,000 tons of inorganic hazardous wastes each year.[82] This prospect was most appealing to Quebec officials, who recognized the absence of sophisticated treatment and disposal facilities in the province and the tendency, as in other parts of Canada and the United States, to dump wastes in a highly inappropriate fashion.

In developing its siting approach, Quebec borrowed some regulatory-style principles in applying various siting criteria to arrive at a list of thirteen potentially acceptable sites. The list was later narrowed to three sites in the Montreal area: the communities of Blainville, Laval, and Mascouche. The Montreal area was particularly attractive for site location because of its status as the province's largest generator of hazardous waste. Although lacking the fuller participatory component so successful elsewhere, Quebec did provide multiple opportunities for local communities to become fully involved once the site selection process progressed. Multiple candidate sites were explored through public hearings and permit application processes, so as not to single out any one community.[83] Moreover, siting would proceed only if a community formally agreed to participate, thus eliminating the threat of regulatory tools such as preemption.

Initial disapproval was evident in all three communities; Laval and Mascouche chose to withdraw from further consideration. The Blainville city council expressed interest in the proposal and passed a September 1980 resolution that invited Quebec officials to explore the matter further in the community of 12,000 residents located forty-six kilometers northwest of Montreal. A host of provincially and locally sponsored citizen groups and public meetings gradually built support in the community. Both provincial and Stablex officials assumed active roles in trying to generate support. Experts on the Stablex Sealosafe process were brought in from Europe and the United States to meet with citizens; trips to a Stablex facility in England were arranged for local media and town leaders; a citizens' advisory committee was formed to meet regularly with provincial and corporation officials; and a cavalcade of public meetings was held in places that ranged from city hall to shopping malls. Opposition groups withered away, unsuccessful in stopping a 1981 referendum to fund an exit to a major highway that was crucial for transporting waste to the proposed Blainville facility. These groups were also unable to mount a serious challenge in mayoral and council elections the following year.

The Quebec approach to public participation was more limited and belated than those employed in Alberta and Manitoba, although it did play an important role in securing the support of Blainville residents. However, unlike Alberta and Manitoba, Quebec was unable to attract multiple candidate sites; it would have lacked a fallback position had Blainville withdrawn. Moreover, the absence of the broader public participation process failed to build the base of trust that was needed and lacking when controversies began to arise concerning facility management in the late 1980s. The broader, more extensive participatory processes of Alberta and Manitoba are more desirable for making siting more genuinely democratic, enhancing the prospects for reaching an agreement, maintaining support for a facility once it begins operation, and increasing the prospects for development of a comprehensive waste management system.

Compensation proved to be a central part of the negotiation process over the facility, although it took a form dramatically different from traditional compensation packages. Blainville is a working-class suburb of Montreal, and it suffered from serious economic problems in the early 1980s, with an unemployment rate above 15 percent at the time of the Stablex agreement. However, it received few of the direct economic enticements offered communities such as Swan Hills or Montcalm. Instead, Blainville was particularly eager to secure a cloverleaf exit that would link it with the main highway to Montreal, as well as an access route to the site.[84] This exit eliminated a twenty-minute bypass through the neighboring city of St. Therese that was widely viewed as a deterrent to economic growth in Blainville. The exit was funded primarily by the Quebec and Canadian governments, although both Blainville and Stablex contributed some funding to the project.

The economic attractiveness of the Stablex facility was further enhanced by a transfer of land from the Canadian military to the city of Blainville for use as an industrial park. More than 2,000 acres at Camp Bouchard went largely unused after World War II, other than a small section where arms were stored. They were well suited for both the Stablex facility and an industrial park, given their proximity to the Blainville work force and the new highway artery that would make Montreal so accessible. Local officials contend that the Stablex facility and the highway exit have facilitated considerable economic growth. The Blainville population grew from 12,000 in 1981 to 20,645 in 1989, and the city has been successful in luring new industrial and residential development. It also opened a major new equestrian park that is integrated with 600 new upper-class residences.

Institutional reforms and innovative policy professionals were not central to the Blainville agreement. Quebec had eschewed the crown corporation approach because of anticipated costs to the province and a perceived lack of provincial expertise in facility construction and management.[85] Public officials played an important intermediary role, more active than in market strategies and less domineering than in regulatory ones. In particular, they helped Stablex representatives win the confidence of Blainville officials and residents through the public participation process. They also provided the funding for the crucial highway exit and access road construction, as well as securing the land necessary for the facility from the federal government. A strong alliance between provincial and Blainville officials of the same political party were essential in these dealings.

However, much of the cooperation generated in the early stages has dwindled in subsequent years. The commitment to sharing the waste burden, so important in other cases, was verbalized but never seriously pursued. Blainville residents were assured that the Stablex facility would be only part of a larger process leading toward a comprehensive approach to hazardous waste management. Subsequent facilities, with supplemental treatment and disposal technologies, were to be constructed, and waste recycling and reduction programs were to be developed. Moreover, although waste trading with other provinces and states was never formally ruled out, provincial officials emphasized that it would be done only on a limited basis and through a quid pro quo arrangement. As in the Alberta, Manitoba, and Greensboro agreements, the primary purpose of the facility was to manage locally generated wastes.

Unlike the Swan Hills and Greensboro cases, where serious efforts have been made to fully implement agreements, and Montcalm, where the likelihood of successful implementation appears very high, Blainville has suffered from serious implementation problems. These have generated growing local resistance to continued operation of the facility and have led many residents to question the commitments made by provincial and Stablex officials in the early 1980s.[86] The Quebec case is beginning to resemble the experience of other, particularly American, communities that accepted major waste treatment or disposal facilities in the 1960s or 1970s and have become increasingly resentful that proper management practices are not being followed and that too many wastes are being imported from long distances.

After the opening of the Blainville facility in 1983, Quebec officials failed to provide the necessary follow-up to assure that provincial waste generators would use the facility, that other aspects of an integrated waste management system would be introduced, or that the public would continue to play a significant role in facility oversight and further policy development. Provincial regulations and basic reporting requirements were delayed and were gingerly enforced once introduced. For example, provincial environmental officials promised in 1981, in the middle of the Stablex siting process, that all firms generating hazardous waste would be required to report which wastes they produced and how they disposed of them. Essential regulations were not announced until mid-1985, with the initial report due in April 1986. Under these regulations, companies could be fined from $5,000 to $50,000 for failure to comply. However, by late 1987 less than a quarter of the 3,500 firms estimated to be required to comply had filed the necessary forms and no fines had been levied.[87] The pace has begun to pick up in recent years, but Quebec continues to lack a clear profile of its waste generators and disposal practices. Most observers contend that Quebec ranks second among provinces in hazardous waste generation, with estimates ranging from 952,000 tons to more than 1.6 million tons a year.

Without pressure to comply with the basic procedures, Quebec waste generators have not been moved to use the sophisticated disposal facilities available in Blainville. They chafed at the prospect of paying $275 to $300 a ton for disposal when they had previously paid much less — if anything — to dispose of the wastes in slipshod fashion and faced little prospect of punishment if they continued to do so. As a result, Stablex has been operating at well under its capacity, and many Quebec waste generators are thought to be continuing to dump wastes illegally into solid waste landfills, water bodies such as the St. Lawrence River, and municipal sewers.

Given the shortfalls, Stablex facility managers have periodically threatened to close their doors because of lack of business. The facility was sold to an American-held firm, Concord Resources, in the late 1980s and has attracted greater volumes of wastes in recent years, making its first profit in 1987. However, its increased business has been secured primarily by heavy reliance on imported wastes from the United States and, to a lesser extent, provinces such as Ontario and New Brunswick. Between 50 and 60 percent of the hazardous wastes that Stablex disposes of each year comes from the United States, compared with less than 15 percent from within Quebec. Overall, Quebec accepts five times as much hazardous waste as it ships south of the border, contrary to earlier provincial promises to restrict such activity.[88] Numerous states send wastes to Blainville; the greatest quantities originate in Massachusetts, New Jersey, and New York. Ironically, these states are proud of their rigorous environmental regulatory records but remain unable to open new disposal facilities and hence continue to rely on exporting to meet Superfund capacity assurance targets. In luring American imports, Stablex places particular emphasis in advertisements on the reduced liability risk that it offers because the province has no counterpart to the American Superfund program.[89]

Continued media revelations of illegal dumping and an August 1988 fire in a poorly monitored warehouse laden with PCBs further suggest that provincial officials have failed to devise a sound regulatory system or to supplement the role that Stablex plays in inorganic wastes. In addition, mounting reports that the Stablex landfill has begun to leak and that the facility has let in more than the trace levels of organics permissible in inorganic wastes accepted for treatment have begun to undermine public confidence in the facility. These developments have contributed to a growing perception that provincial officials chose to ignore crucial regulatory matters once the Stablex permits were signed and the facility opened for business. As a result, concerns over environmental and public health problems related to hazardous waste are as high, if not higher, in Quebec than before the facility was opened. In response, Quebec officials have begun to turn to Alberta and Manitoba for ideas to guide their policy reform efforts.

The Quebec experience is in direct contrast to other siting cases, indicating that the approach necessary for successful siting must also be sustained through implementation, to maintain public support and assure development of an effective waste management system. On the one hand, Swan Hills residents clearly continue to view their decision to host a facility as a good one, having been given no reason to doubt the commitments made to them at the time of siting. Greensboro residents feel similarly about their decision, an island of siting cooperation surrounded by enduring conflict and inertia over most other aspects of North Carolina facility siting and hazardous waste management. Manitoba and Minnesota can look to the future with some confidence in the systems of hazardous waste management that they have developed. None is foolproof, but all reflect a reasonable effort at reform through a voluntary process that emphasized burden sharing.

On the other hand, Quebec is painfully reminiscent of those cases employing the regulatory and market approaches to siting. Failure to provide a comprehensive system of facility siting and waste management or to honor promises once they have been made only delay reaching the point at which cooperation and effective policy become possible. Quebec, despite its promising siting agreement in the early 1980s, joins the roster of states and provinces in which distrust dominates and uncertainty over future policy direction prevail.

5

Low-Level Radioactive Waste

THE EMERGENCE of siting agreements is attributable to the new voluntary, burden-sharing approach, not to unique circumstances. The common characteristics of successful siting could prove useful in other areas of waste management in which the Nimby syndrome prevails and siting has become contentious. Few cases could provide as tough a test of the replicability of the Alberta approach as low-level radioactive wastes (LLRW). Solid wastes pose certain environmental risks, but neither they nor most hazardous or biomedical wastes are a proven danger to both human health and the environment through contamination comparable to radioactive wastes. In Canada and the United States, the siting of facilities for either high-level or low-level radioactive wastes has been controversial.[1] The radioactive waste issue has probably triggered more political conflict over siting than hazardous waste. Consequently, neither Canada nor the United States has opened a new storage or disposal facility for either high- or low-level waste in more than twenty years. Meanwhile, the waste continues to be generated, being either stored on-site or shipped to one of the few remaining waste disposal facilities that were opened before controversy arose.

Not surprisingly, Canada and the United States have in recent years pursued approaches to siting similar to those that have prevailed in hazardous

This chapter was coauthored with William C. Gunderson and Peter T. Harbage.

waste. Regulatory approaches have been common as either federal, state, or provincial authorities have attempted to impose a facility on some community with a minimum of prior consultation. As the growing scholarly literature on this area confirms, the familiar Nimby reaction ensues and the search for new facility sites begins anew.[2]

While high-level radioactive waste siting remains hopelessly deadlocked in both nations, some promising possibilities exist for cooperation in low-level waste siting, particularly in the province of Ontario, which otherwise has experienced classic Nimbyism in both hazardous waste and solid waste facility siting.[3] Contrary to virtually all prior LLRW siting experience in Canada and the United States, multiple communities in Ontario are voluntarily and actively exploring the possibility of hosting a facility. Moreover, they have agreed to do so through a process that draws heavily on the voluntary approach to siting initiated in Alberta. The Ontario case indicates that the Alberta approach to siting may have replication prospects beyond the area of hazardous waste facility siting.

Early Stages of LLRW Management in Canada and the United States

The technical and political dimensions of low-level radioactive waste management may be more similar between Canada and the United States than they are for hazardous waste. Both federal governments played a fundamental role in endorsing and subsidizing the nuclear power industry in the decades following World War II. Despite tendencies toward regulatory decentralization, particularly in Canada, both federal governments have assumed central responsibility for radioactive waste management. They continue to play a far more central role in virtually every aspect of radioactive waste regulation and management than they do for hazardous, biomedical, or solid wastes. Both governments have delegated more authority to provinces or states in this area in recent years, although this trend has advanced more rapidly in the United States than in Canada.

Canada and the United States also speak much of the same technical language concerning nuclear power and radioactive waste. The Canadian nuclear industry continues to promote actively its compact, Candu 3 nuclear reactor in the United States.[4] Moreover, waste definitions and classification schemes are similar, and the technological options for waste reduction and disposal are largely identical. Perhaps most important, both nations have experienced strikingly similar patterns of conflict for radioactive waste facility siting. The seemingly intractable conflicts evident in the American low-level radioactive waste cases closely resemble the hazardous waste facility siting efforts using the regulatory and market approaches. Canadian low-level radioactive waste facility siting until 1986 was in many respects a carbon copy of the approach — leading to Nimby results — that continues to be used in many individual states and multistate compacts.

The major sources of commercial LLRW in both nations include nuclear power plants, biomedical and industrial research, and nonmilitary governmental projects. The most common source remains nuclear power plants, which, in the United States, generate 51 percent of the nation's LLRW when measured in cubic feet, 81 percent when measured in curies.[5] Many of these plants were opened in the 1960s and are nearing the end of their functioning capacity. When the plants are closed, they no longer will generate new LLRW but must be decommissioned, a complex and expensive process that will require removal of radioactive and nonradioactive materials.

The Pacific Ocean and the Atlantic Ocean served as repositories for American LLRW during its first three decades of generation. Commercial land disposal of these wastes began in the early 1960s, with the opening of six facilities that used shallow land burial technology. The process was primitive: burying waste in trenches and often storing waste in cardboard boxes until the burial was completed. Three facilities were closed during the 1970s because of leakage and other forms of contamination. The Beatty, Nevada, facility closed in January 1993, and the two remaining facilities, located in Washington and South Carolina, are reaching the end of their anticipated operational lives. The facility shortage has triggered new legislation, the Low-Level Radioactive Waste Policy Act of 1980 and its 1985 amendments, which turned considerable siting authority over to the states and set up a complex system of incentives and timetables for states to assume responsibility for their wastes.[6] However, most states or compacts have employed traditional regulatory or market siting strategies, and not a single new facility has opened since the new legislation was enacted.

In Canada, the Atomic Energy Act of 1946 established the basic framework for handling radioactive waste. The act created the Atomic Energy Control Board (AECB), which retains authority for licensing and regulating the nuclear industry, including waste storage and disposal. It is distinct from Atomic Energy of Canada Limited (AECL), which is responsible for federal research and development related to nuclear energy, and Ontario Hydro, a major provincial public corporation that is responsible for management of Ontario's nuclear power plants and is the monopoly power producer in the province. The AECB shares regulatory authority in that it typically confers in a working group process with related federal and provincial agencies. Provincial involvement is a particular concern because Canadian LLRW is concentrated within its largest province. Sixteen of the nation's eighteen nuclear power plants are located in Ontario and, by most estimates, more than 90 percent of Canada's LLRW is generated by Ontario, forcing the federal government to be sensitive to provincial concerns.

While the United States has depended upon ocean dumping and later commercial facility development, Canada has generally favored a quiet, de facto approach to LLRW, encouraging storage and disposal at or near the sites where the waste is generated. Most nuclear power plants keep wastes on-site under interim storage provisions, just as American plants have begun to do as disposal options narrow. In Canada, however, many of these wastes are expected to be shipped to one or more permanent facilities. Wastes generated from the use of radioisotopes sold by the AECL Commercial Products Division must be returned for burial at the AECL Chalk River Nuclear Laboratories in central Ontario. Finally, serious waste contamination problems arose in and around the town of Port Hope, which put the radioactive waste disposal issue on the Canadian political agenda nearly twenty years ago and triggered a search for more effective siting and disposal methods.

The Evolution of Canadian LLRW Management

The problem of radioactive waste disposal at Port Hope preceded the formation of the AECB and the development of the nuclear power industry by more than a decade. Wastes generated by various domestic and military activities were deposited at several sites within the Town of Port Hope from 1933 to 1948. Some of these wastes were transferred to nearby locations, including Welcome and Port Granby, in later years.

However, radioactive contamination was not discovered in the Town of Port Hope until 1975. This contamination, mostly in the form of radioactive soil, had been widely dispersed around the community. Radioactive waste was deposited in the Port Hope harbor, and a number of buildings in the town were constructed with contaminated material. As a result, the AECB tore down sections of the town between 1977 and 1979, once the extent of contamination was realized. Approximately 5,000 truckloads of waste, comprising 200,000 cubic meters, were transferred to Chalk River. Other parts of the community remain fenced off as nuclear reservations because of their high levels of contamination.[7]

After the AECB completed its remedial actions in Port Huron and other contaminated areas, it asked Eldorado Resources Limited, a federal crown corporation now known as CAMECO, in 1980 to begin work on the design and siting of advanced LLRW disposal facilities. However, Eldorado's approach to siting resembled Ontario's approach to hazardous and solid waste facility siting and that of troubled LLRW compact states such as Connecticut, Michigan, and New York. In the absence of public consultation, Eldorado began the process of examining alternative sites, using various technical criteria to assess the suitability of each potential area. In August, authorities announced that Eldorado "had taken options on two properties in the Port Hope area" and intended to complete surficial geological assessment on both sites before selecting one.[8]

Typical of regulatory approaches, the first opportunities for public participation were to be delayed until after the siting announcements and were to be confined to the federal environmental assessment process. An environmental assessment panel was appointed and received the formal site proposal and an environmental impact statement. This material was also made available to the public, and local issue identification, or scoping, meetings were planned to acquire public input. However, the public reaction to learning of the two site candidates was so strong that the meetings were postponed until after a final site had been selected. Public opposition in response so intensified that the federal government suspended the process and dropped the two sites from further consideration.

The Emergence of the Voluntary Process

The minister of state for forestry and mines appointed an independent Siting Process Task Force in December 1986 to develop a less confrontational site selection process. The task force was commissioned to focus its efforts on "the development of a process for siting a disposal facility in Ontario for the existing, ongoing and historic wastes in the Port Hope area, and where advantageous, for the disposal of other existing and ongoing low-level wastes located in the province. The main objective of the process was to be the voluntary identification of one or more host communities, each with a suitable disposal technology." [9]

The task force suggested that a process driven by social instead of technical criteria be applied to resolve the low-level radioactive waste disposal problem.

The successful experience of siting a comprehensive hazardous waste disposal facility in Alberta became a model for the task force as it rethought the process of LLRW disposal facility siting. The new approach, proposed in a 1987 task force report entitled *Opting for Cooperation*, drew heavily on the Alberta hazardous waste process and, in the words of one task force member, "built on the mistakes [Alberta] made." The task force was highly critical of the AECB for allowing technical criteria to drive the previous selection process, failing to consult with affected communities until site selection was effectively completed, and insisting that low-level radioactive waste be disposed of through land burial, not stored or handled by other methods.

The task force strongly recommended that a new siting process be established that allowed potential volunteer communities the opportunity to learn about the problems of LLRW management as well as the possible payoffs if they decided to accept a site. The task force's position was that volunteer communities should be free to opt out at any time in the exploratory phases of the process. Those deciding to proceed would become full partners in making decisions on all relevant matters. As a result, site selection and design, type of disposal technology, and impact management, among other important features, were left open to negotiation instead of being imposed by authorities. These provisions were not only a dramatic departure from prior LLRW siting efforts in Canada and the United States, but they also featured many of the common characteristics for successful siting evident in hazardous waste.

Implementing the Voluntary Process

The 1987 release of the task force report was followed by efforts to refine and ultimately implement the new siting process. On September 30, 1988, the minister of energy, mines, and resources appointed a new Siting Task Force (STF) to implement the first phases of the voluntary approach to siting. A second task force was asked to report back to the minister in eighteen months regarding potential volunteer communities, a tentative discussion of disposal and storage options, the terms of reference for negotiation with local communities, and detailed cost estimates of implementation. Because of the high levels of interest generated by the process, the new STF was given twenty-three months to complete its work.[10]

Besides trying to clarify a few matters concerning site suitability and alternative technical options for potential host communities, the STF had to address some elements of the consultative process, particularly as they concern

community liaison groups (CLGs). First, the STF decided that all CLG meetings should be open to the public, despite the absence of any federal or provincial regulations that mandated openness. Second, depending upon the
response of those attending the meetings, a CLG could decide to opt out of
the process at any time. Furthermore, a community could decide to receive
newly generated, as well as historic, radioactive wastes, even if the amounts
would be relatively small and thus would constitute only part of the larger
waste management solution.

Formal consultations with regions and communities began in November
1988, as exploratory letters were sent to 850 communities. Four months later,
follow-up letters were sent to invite two representatives from each community
to attend one of eight regional information meetings. At these meetings, community representatives were told of the need for improved management of
LLRW, the process principles already established, and some of the safeguards
available to those interested in continued participation. Community representatives asked about how communities would be consulted; how they would be
defined in the first place (some contiguous communities wanted to proceed
together); and how their CLGs, if formed, would measure community acceptance. They also asked about the characteristics of and the hazards associated
with LLRW, the dangers involved in transporting LLRW to their communities, the ongoing problems associated with managing a facility, the type and
number of jobs available to local citizens if a facility were built, and the impact on tourism.

After the regional consultations, twenty-six communities requested additional meetings or information or both. Between the regional meetings and
follow-up informational sessions held by the STF, five communities dropped
out. After the informational sessions an additional seven withdrew, leaving
fourteen to enter the next phase of the process. The remaining communities
then formed their CLGs. Various formal and informal contacts, including ads
in local newspapers, were used to compile tentative lists of candidates to be
CLG members. The STF staff called those listed to see if they met the criteria.
A final list was printed in newspapers to allow a two-week comment period
regarding the representativeness of it. By the end of November 1989, ten
CLGs were formed. An additional one was completed in January 1990, followed by one in March and the last two in April. By request of the local councils, the Siting Task Force had taken responsibility for selecting CLG
members.

As this process unfolded, the advantages of a voluntary over a regulatory
process became clear. In the voluntary process with which the STF has experi-

mented, the typical kind of Nimby confrontation has not occurred. To the contrary, the various CLGs networked: they shared information and ideas and learned from each other's experience, as exemplified by the January 1990 Thunder Bay meeting of chairs and facilitators of community liaison groups.

The results of extensive consultation within communities — between the communities and the STF, and between the communities and outside experts selected by the communities — varied considerably. In Atikokan, Ear Falls, Manitouwadge, Mattice-Val Cote, Red Lake, and Upsala, both the CLGs and the local councils rejected further consideration of hosting a site. The CLGs of the United Townships of Buchanan, McKay, Rolph, and Wylie and of the Townships of Clara, Head, and Maria declined to continue. The CLGs of Chalk River, Deep River, and Hornepayne chose to withdraw, but their re-spective councils approved movement into the next phase. The Elliot Lake and James Township CLGs wanted to continue, but their respective councils refused.

The concerns of those communities that opted out were familiar: the po-tential health risks, and the risks to ground and surface water, were perceived to be too high, with a baseline health study considered absolutely essential before proceeding; the potential negative effects on economic development and tourism were believed to be too severe; the potential transportation risks and effects were also thought to be too great; and the long-term integrity of technology to manage LLRW was feared to be too uncertain.

Other concerns involved a lack of trust in the federal government's willing-ness to pay for compensation and a well-managed facility, as well as a continu-ing wariness of the AECB. The AECB lacked the public credibility of crown corporations developed for hazardous waste in Alberta and Manitoba, in large part because of its history of closed operations and suspect waste management practices. Communities also worried that if they were to accept LLRW, other sorts of waste, including high-level radioactive materials or hazardous wastes, would ultimately migrate there for disposal.[11] Fears were exacerbated by On-tario's controversial efforts to locate sites for disposal of hazardous and solid wastes in the 1980s and early 1990s.[12] In both of these cases, technical, top-down approaches to facility siting have led to the familiar outcome of gridlock.

Possible Site Volunteers

The large number of withdrawals from consideration did not constitute the end of the voluntary siting process. Despite the concerns that prompted a number of CLGs and councils to opt out, several communities remained interested and continued deliberations with STF officials through 1994. More-over, one source community was interested in cleaning up LLRW located within its own boundaries.

Among the potential volunteer communities, the Deep River Council (a neighbor of the Chalk River facilities of AECL) chose to proceed after its CLG said no for four cluster communities. These communities are located in northeastern Ontario, with Ottawa 115 kilometers to the south, and have a combined population of approximately 3,800. The two major concerns of the citizenry were that the historic wastes of the Port Hope area be managed there before being moved anywhere and that an accident or spill could occur if LLRW were moved through surrounding areas. The Deep River Council agreed to continue the process, if, for any reason, Port Hope was deemed an unsuitable site; any waste relocated to Deep River is transported by rail with the material stored in suitable containers; and during or at the end of phase four — when the final siting selection is made — a referendum is held to deter-mine whether or not to continue the process, with any associated costs to be paid for by the federal government. Chalk River agreed to go along with the Deep River Council. Because no sites are located in Chalk River, it seems to want to be part of the process to protect its interests.

Geraldton, a community of 2,882 located the furthest west of all the poten-tial sites, approximately forty-five kilometers northeast of Thunder Bay, was intrigued by the process, given its potential contributions to economic diversi-fication. At the same time, the community articulated concerns about nega-tive affects on tourism and health, with the latter leading to a request for a baseline study to make possible the monitoring of health impacts. Uncertainty also existed over Geraldton's ability to annex a parcel of land intended for any disposal facility. Other issues of debate involved arsenic content in the waste, transportation, suitable long-term management, and the amount of equity compensation the community would receive for accepting LLRW. These con-cerns, combined with procedural difficulties in completing technical work before a planned November 1994 referendum, led to Geraldton's decision to withdraw from the process.

In addition to the potential volunteer communities, some source commu-nities have decided to proceed to the next phase. Port Hope, located forty-five kilometers east of Toronto with a population of 10,300, has been most active among communities in this category. Its citizens wanted, most of all, to find a permanent site for the wastes deposited within town boundaries. Further-more, they want to play a major role in the cleanup and hope to do this by having their CLG stay in operation until the process is over. To fulfill its role, the CLG seeks to be empowered to hire consultants and contract for studies as necessary. In short, the Port Hope community, through its CLG, wants to participate in the establishment of criteria for selecting a site, in the cleanup of old sites, and in the decision regarding which wastes will be moved from the old to new sites. During the latter months of 1992 the Port Hope community demonstrated intensified interest in the possibility of hosting a central facility. In October 1992 the Port Hope Council passed a resolution expressing a willingness to continue negotiations but emphasizing key conditions such as components of the compensation package.

Construction of a new harbor for Port Hope has become central to deliberations over LLRW facility siting. The International Joint Commission has identified Port Hope Harbor as among the forty-three most environmentally hazardous sites bordering the Great Lakes and in need of immediate "remedial action" because of extreme contamination from LLRW and numerous other toxic substances.[13] In response, the Port Hope CLG recommended that Port Hope Harbor be included under the criteria to be developed jointly by the Port Hope CLG and the Siting Task Force. It also recommended that the government of Canada consider creation of a new harbor in Port Hope as a demonstration of its commitment to the environmental integrity of the Great Lakes. LLRW management thus could be formally integrated with the larger issues of sediment contamination and areawide remediation, possibly leading to increased public support and greater integration of environmental management in this part of Ontario.[14]

Hope Township has also remained active as a source community. Its CLG stipulated that it wants control of how wastes are evacuated and transported from the Welcome township's LLRW site, as it is very concerned about radioactive and arsenic wastes. Because the township will be affected by decisions regarding what to do with its historic wastes, the CLG wants it to be part of the consultations that determine where the new site will be located. And the CLG wants to continue to function until the process is complete. The Hope Township Council went on record in support of the CLG.

Newcastle, a Toronto exurb with 34,100 residents, has continued its participation in the process. Like other source communities, its involvement has been motivated in large part by concern over Port Granby wastes. Finally, Scarborough, a city of 485,000 located just east of Toronto, did not form a CLG because the disposal of a small amount of LLRW in that community has already been worked out by formal agreement between it and the federal and provincial governments.

As this winnowing among possible candidates continued, a new task force was created in 1992 and was expected to complete siting deliberations over a four-year period. The volunteer community will explore the technical, environmental, and social effects of receiving wastes while source communities will look at the effect of removing wastes. Joint decisionmaking will be the key and will include the beginning of operational planning in the remaining volunteer community (Deep River) and the three remaining source communities (Port Hope, Hope Township, and Newcastle).

The voluntary process has not been derailed by two potentially disruptive changes in government. In 1989 the New Democratic party, strongly opposed to expansion of nuclear power, replaced the Liberal party in governing Ontario. The Liberals replaced the long-ruling Progressive Conservatives after the 1993 federal elections. As in Manitoba hazardous waste facility siting, the changes have had no discernible impact on the LLRW siting process.

The signs of evolving cooperation in LLRW siting in Canada are remarkable, given the historic opposition to siting all sorts of LLRW disposal facilities in both Canada and the United States. Even more than in hazardous waste policy, examples of LLRW siting successes are hard to find in either nation. The dismal American experience in LLRW siting is one that Canada has learned from and chosen to avoid. Examining the poor track record of many of its own provinces in hazardous and solid waste facility siting, and its highly contentious early approach to LLRW siting, Canada undertook a significant reform of its siting policy in 1986. As in the successful cases involving hazardous waste facility siting, several common design characteristics were found.

First, early, extensive public education and participation efforts were crucial in Ontario. Public information and participation activities were not as numerous as in the western provinces but were extensive and were based on the principles of voluntarism. All involved communities were given the option to withdraw at any point in the process. They were also assured of direct input in the selection of technology at any facility that might be opened.

Second, Canada's LLRW siting efforts have promoted the notion of burden sharing in a variety of ways. Perhaps most important, siting officials have remained open on the ultimate number of places that will participate in waste management and whether future sites will provide waste storage or more permanent disposal. For example, one major facility could emerge, with other communities developing more modest facilities. Such options have been foreclosed in the American states or compacts, where an intensive search usually is made for a single site to employ a single disposal technology to dispose of all the wastes generated within that geographic area.

The Canadian process has also benefited from a general sense of assurance that one facility will not become a magnet for wastes from other parts of Canada. Because Ontario generates the vast majority of Canada's LLRW, little threat exists of exploitation by neighboring provinces. Moreover, officials have stated that any new LLRW facility will not accept waste from the United States or other nations. A shift in position could undermine the cooperation that has emerged thus far.

The final aspect of burden sharing, combining siting with expanded efforts to reduce the volume of waste being generated, has proven less prominent in LLRW deliberations in either Canada or the United States than in the hazardous waste facility siting agreements in Alberta and Manitoba. Waste generators in both nations have made significant strides in waste reduction in recent years, but this development has not formally been incorporated into siting deliberations. By contrast, several European nations have made this linkage central to their siting policies.[15]

Third, Canadian LLRW siting officials have proven open to alternative siting technologies and varying types of compensation packages. In Port Hope, for example, the issue of facility siting has been directly linked with harbor dredging and cleanup, an action that, if completed, would offer considerable economic and environmental benefits for local residents.

Fourth, officials have been successful in devising new managerial partnerships to defuse concern over long-term commitments to the technical and financial feasibility of new facilities. As in hazardous waste, many private and public LLRW management corporations in both Canada and the United States have shoddy records of maintaining high levels of operational proficiency and assuring fiscal commitment to safe facility operation. Slipshod public and private management helped undermine an American case in which, under a voluntary approach, an agreement may have been possible. By contrast, the Canadian Siting Task Force has proven successful at winning public trust and facilitating genuine dialogue over siting options in multiple communities. It has helped overcome understandable public reservations about both Eldorado and the AECB, a pair of public entities with long and somewhat controversial records.

These four broad features — public participation, burden sharing, compensation and protection, and a credible governmental role in siting and facility management — appear to have contributed to the evolution of cooperation in Canadian LLRW management, much as they have in hazardous waste facility siting agreements.

The Evolution of American LLRW Management

The 1980 Low-Level Radioactive Waste Policy Act and its 1985 amendments represent a significant departure from traditional American approaches
to LLRW that have been dominated by the federal government. The laws featured a series of federal mandates and incentives that encouraged individual
states to form multistate compacts to manage LLRW.[16] The legislation retains
federal authority for setting safety standards and offers a system of timetables,
incentives, and penalties related to compact creation. As with many other interstate compacts, the radioactive waste compacts must be approved by Congress before becoming official.

The legislation also delegates a significant degree of subnational authority.
It leaves to individual states the decisions concerning the other states (if any)
with which they will share compact responsibilities, how they will design and
implement their siting processes, and what methods they will use for disposal.
Among the powerful incentives that the federal government provides to states
to enter into compacts instead of managing waste on a solo basis is that those
states that belong to a congressionally approved compact have the authority
to exclude wastes from other states.

The compact approach represents a unique federal government effort to
delegate regulatory authority to individual states or clusters of them. Its enactment was driven in large part by the efforts of the governors of Nevada, South
Carolina, and Washington. Their respective states hosted the only three waste
disposal sites that continued to operate in the nation after three other sites
closed in the late 1970s. The governors' threat to restrict access to their states'
sites prodded Congress to accept the compact approach that was developed
by the National Governors Association.[17]

The success of the compacts was to be measured by their ability to meet a
series of federally established milestones that anticipated a fully functioning
system of sites by 1993. The timetables were delayed and altered in the 1985
amendments, when it became obvious that the initial legislation had been too
ambitious. The current deadlines are outlined in table 5-1, although select
portions were invalidated by a 1992 U.S. Supreme Court decision that
deemed unconstitutional the requirement that states "take title" to all wastes
after specified deadlines.[18] However, much of the remaining structure was not
overturned, and, at least in some cases, the siting process continues.

All of these complicated factors were designed to promote policy coordination and facilitate long-term support for waste disposal agreements. They operated on the assumption that states and compact regions would perceive long-

Table 5-1. *Low-Level Radioactive Waste (LLRW) Policy Amendments Act Deadlines*

Deadline	Provision
January 1, 1986	Each state to have joined a compact or to have enacted legislation to develop its own site; surcharge not to exceed $10 per cubic foot.
July 1, 1986	Generators in states that did not meet January 1, 1986, deadline are subject to double surcharges until December 31, 1986.
January 1, 1987	Generators in states that did not meet the January 1, 1986, deadline may be denied access to operating disposal sites.
January 1, 1988	Compacts to have named host states; unaligned states to have developed a siting plan and schedule and to have delegated authority for development; surcharge not to exceed $20 per cubic foot; noncompliance states subject to doubled surcharges.
July 1, 1988	Noncompliance states subject to quadrupled surcharges.
January 1, 1989	Generators in states and compacts that did not meet the January 1, 1988, deadline to be denied access to operating disposal sites.
January 1, 1990	Compacts and unaligned states to file a complete operating license application; letter from governor stating that the unaligned state will have provisions for LLRW disposal in place by December 31, 1992, may be submitted in lieu of application; surcharge not to exceed $40 per cubic foot; failure to comply may result in denial of access to operating disposal sites.
January 1, 1992	All compacts and unaligned states to file operating license applications; letter from governor no longer sufficient for compliance status.
January 1, 1993	Sited compacts to be empowered to restrict import of non-compact LLRW.
January 1, 1996	Surcharge rebates cease; states without disposal capacity upon request must take title to wastes produced by in-state generators. (Since a 1992 Supreme Court decision, states are no longer required to accept title to in-state wastes.)

Sources: Low-Level Radioactive Waste Policy Amendments Act of 1985, 99 Stat. 1842 (January 15, 1985); and *New York* v. *United States*, 1125 S. Ct. 2408 (1992).

run advantages from entering into compacts and accepting surcharge rebates and other compensatory benefits. In this process, it was assumed, they would successfully override political resistance to facility siting. The approach hinged, in many respects, on a market-type presumption that compensation would serve as a reliable lubricant of siting conflict. In practice, however, many individual states and compacts have relied on a regulatory approach to siting, one that delays discussion of compensation until after authorities announce their site selection decisions.

The compact approach has enjoyed one degree of success: forty-two states have entered into a total of nine congressionally approved compacts. The fifteen states that belong to the two compacts that feature disposal facilities in continued operation (in South Carolina and Washington) are set for the near future, having secure facility access and the ability to exclude out-of-compact wastes if they so choose. For the remaining states and territories, however, the compact approach has largely been a nightmare. No new siting agreements have been reached in any state or compact, and none is in sight in the near future. South Carolina has allowed a small amount of LLRW imports from states demonstrating some continued commitment to resolving their own waste disposal problems, imposing an increasingly steep surcharge in the process. No assurance has been made, however, as to how long this export option will remain available, as the state is free to close it at any time. Those states perceived as not making a reasonable commitment to waste management have been completely cut off from further access to the facility. Consequently, many states are forced to store LLRW at the point of generation; California, for example, has more than 2,250 licensed users of radioactive materials, not including nuclear power plants, and all of these must scramble to secure storage capacity. Many of the newly formed storage sites were never intended for long-term storage or disposal, and they pose significant safety and security problems. Increasingly, LLRW that cannot be shipped to South Carolina or Washington must be stored in this fashion, either at the site of nuclear power plants, hospitals, medical research institutes, or at other facilities that generate some amount of LLRW.

The siting approach most commonly used in the states and compacts has eschewed broad public participation and comprehensive waste management in favor of a traditional regulatory approach driven by technical criteria. In Connecticut, for example, the state's Hazardous Waste Management Service used such criteria to select three towns as its preferred candidates for a facility to manage LLRW from Connecticut and its Northeast Compact partner New Jersey. No prior consultation was made with these communities, and no prior warning was given that they would be finalists. As a first selectman of one of these areas explained after learning the news, "I'll never forget it. I was in shock. I had to sit down for a minute."

Residents of the other two potential host communities responded similarly, but then wasted little time in taking vigorous collective action against the proposal. Public meetings were finally held, but these became rallying events for opponents. Several interviewees recalled that, at one meeting, an official of the Hazardous Waste Management Service said, in essence, that "the site will go in one of these areas, and you cannot stop it." However, after months of Nimby opposition, the proposals were withdrawn, leaving the siting process in Connecticut and New Jersey up for grabs. LLRW is no small issue in these states, as combined they produce approximately 7 percent of the nation's total new LLRW generated each year.

Michigan exercised similar diplomacy and participatory skills in advancing its siting plans. The state grudgingly agreed to become the siting host for the seven-state Midwest Compact in 1986 and received considerable financial support through the rebates made available to the compact from the federal government. But like Connecticut, Michigan relied on technical criteria in the absence of public participation in its process of selecting three sites. In October 1989 three rural areas scattered about the states were informed, by mail, that they were the finalists. None was aware that it was under consideration for managing LLRW from around the midwest region. As one town supervisor recalled, "I first found out when the press called to ask me how I felt about all of this. . . . It was really upsetting to find out that way." From this point, the Michigan siting process paralleled the Connecticut experience. After nearly two years of contentious public meetings and Nimby opposition, the state withdrew the three sites from further consideration. However, having failed to develop any alternative siting process, it was ousted from continued participation in the Midwest Compact, left on its own to find a siting solution, and cut off from any further access to the South Carolina facility.

Other compacts and states have struggled in similar fashion, although a few have tried to incorporate some of the conditions that have proven crucial to the cooperation emerging over Ontario LLRW siting and in the earlier examples of successful hazardous waste facility siting. Among these, the Central Compact and its host state, Nebraska, remain the closest American counterpart to this alternative approach. It emphasized a number of the key themes raised in the Ontario case and made far more progress toward a cooperative solution than other states and compacts, although its siting process faced some serious problems and appears to have stalled, perhaps permanently.

An Attempt at Voluntarism

Between 1981 and 1987 Nebraska followed much the same process as had Connecticut. The state was negotiating its status in a compact, and in March

1982 the Central Compact was formed. In addition to Nebraska, the compact was joined by Arkansas, Kansas, Louisiana, and Oklahoma, which collectively generate about 9 percent of the nation's LLRW each year. Then, beginning in 1986, the Nebraska Department of Environmental Control drafted several regulations regarding the construction of a LLRW facility. Several public hearings were held on the compact and the facility regulations, but few citizens attended the meetings. In June 1987 the compact selected a private firm, US Ecology, to develop, construct, and operate the Central Compact facility.

At this point, the Nebraska case began to diverge from the one in Connecticut. Nebraska was chosen as the compact's host state in 1988. This stimulated a citizens' group, Concerned Citizens for Nebraska, to gather the more than 150,000 signatures necessary to hold an initiative vote on whether or not the state should remain in the compact. In November 1988, 69 percent of participating voters supported continued compact involvement. The initiative (known as Initiative 402) proved a major statewide issue, and several groups on both sides of the question organized and advanced their respective cases. The questions raised during the campaign focused not on site-specific issues, but on larger concerns, such as whether or not Nebraska's waste generators would have to shut down if Nebraska rejected the possibility of hosting a site.[19]

The initiative campaign overlapped with the start of Nebraska's public participation efforts in April 1988. The Nebraska Citizens Advisory Committee (NCAC) and the Nebraska League of Women Voters played important roles, supplementing related efforts of US Ecology. The NCAC was established by the compact and was designed to reflect diverse constituencies, as it included representatives of farming, ranching, business, environmental, and natural resource interests. It was intended to gather public opinion and to advise US Ecology on key siting and facility design issues. Over a seven-month period, the committee held six public meetings, at which the public was encouraged to give its input on any aspect of the siting program. At least one meeting was held in each major region of the state.

The League of Women Voters sponsored two series of public workshops in 1988. The first series, held in May, involved US Ecology presentations of the siting criteria that were going to be used. Citizens were encouraged to comment on the appropriateness of the proposed criteria. The second series featured a US Ecology presentation of the proposed facility design and, again, sought public suggestions. Five workshops were held in each of these two series, generating thousands of comments and suggestions from all around the state.

US Ecology played the most important role in the public participation process. As these other efforts were taking place, it mailed out invitations to all Nebraska municipal-level governments and asked them if they were interested in the possibility of hosting a facility. Using the same voluntary philosophy successful in other siting cases, the firm made clear that it would not enter an area for site testing unless invited by local authorities. Moreover, an initial positive response was not binding; communities were told that they could withdraw whenever they wished.

In response, fifty-two municipal governments and twenty counties expressed interest in exploring the possibility of hosting a facility. Within these fifty-two communities, 111 potential siting areas (PSAs) existed based on prior research by US Ecology. The response was astounding, given the pattern of near-immediate rejection of siting proposals in virtually all other states and compacts. As the process continued, some candidates withdrew, either for political or geological reasons, but 27 PSAs in eleven counties remained involved. The pattern was similar, in many respects, to the winnowing that occurred among potential volunteer sites in hazardous waste in Alberta and Manitoba and is occurring in LLRW in Ontario.

The process then began to sour. Instead of pursuing continued public dialogue with all possible candidates, as in Ontario, the Central Compact moved quickly to narrow the field and attempt to reach an agreement. The NCAC used a blind format involving technical information on remaining areas to select three for further examination and eliminate all other possible candidates. The NCAC chose rural sites in Boyd, Namaha, and Nuckolls Counties, and US Ecology designated specific areas within these counties as its finalists on January 18, 1989. This was little more than three months after the statewide referendum, reflecting the fast pace of site selection after the initial public participation efforts. Hence, little opportunity existed for the sorts of follow-up discussions and broad review of multiple candidates that were crucial to the other siting agreements.

Generous compensation packages were offered, providing a steady flow of new, largely unrestricted income to the three county finalists. In 1989, for example, each of the three final counties received $100,000. US Ecology focused most of its subsequent analysis on Boyd County, in particular a site in McCully Township (population 160). This township is located near the village of Butte (population 540), upon which it is dependent for all educational and fire protection services. US Ecology was attracted to the site for technical reasons as well as for strong support the facility had from the Butte Board of Commissioners, which had twice passed resolutions supportive of hosting the facility. US Ecology announced on December 31, 1989, that the McCully

Township site had been selected as the preferred host area, closing its information office and ending related efforts in Nuckolls and Nemaha Counties.

The selection of McCully was a serious blunder that undermined the public trust and dialogue that had developed. Despite the support of Butte elected officials, numerous warning signs were evident that any attempt to impose a siting decision in Boyd County would be ill fated. One day before the January 1989 announcement that a Boyd County site would be one of three finalists, the Boyd County commissioners had voted to rescind their invitation to US Ecology. Neighboring communities of Spencer (population 645, located twelve miles east of Butte) and Naper (population 171, located twelve miles west of Butte) had registered strong opposition in advance of the January announcement. Butte, Spencer, and Naper have a long-standing history of animosity toward one another, with a series of skirmishes over the placement of railroads, highways, the county fair, and the county seat. Some of these battles dated back to the nineteenth century, but the tensions endured through the 1990s, enflamed further by the LLRW proposal. Moreover, the communities sparred over a proposed merger of the Butte and Spencer High Schools, which would mean that one community would lose its secondary school. Resentment also existed over division of the compensation package; Butte received $300,000 in virtually unrestricted funds for each of several past years, while its neighbors received nothing. Finally, a strong grass-roots organization, Save Boyd County, had already formed to oppose the facility before US Ecology chose to press ahead and would later seek advice from national antinuclear groups on civil disobedience and harassment tactics.[20] In short, Boyd County was not an auspicious place to pursue siting, especially when opposition was already mounting and other volunteer candidates had expressed interest in the possibility of hosting a facility.

The selection of Butte was understandably viewed by many as a fundamental breach of US Ecology's earlier promise to pursue siting only where a genuine volunteer commitment was evident. As late as December 1988 a US Ecology vice president had reassured the Boyd County commissioners that the firm would leave if at any future point local citizens wanted them to do so. Despite the strong opposition, a group of Butte citizens, People for Progress, have continued to rally support for the site and claim to have 500 supporters throughout the county. Since the siting process began, two Butte Board of Commissioners elections have been held, and on both occasions candidates supportive of the site have been elected or reelected. Butte's elected county executive has also continued to support the siting process. However, all subsequent US Ecology efforts to hold public meetings and discussions of the proposal have proven divisive, leaving the future of the siting proposal for Butte in disarray. The credibility of US Ecology has been badly damaged, in Boyd County and the remainder of Nebraska, by this episode, making it an unlikely candidate to lead any resumption of the voluntary siting process. "I don't think that US Ecology could hold another meeting in [Boyd County]," said one siting opponent. "They would just get run out on a rail."

This failure should not, however, mar the considerable achievements of the Nebraska siting process. More than in any other compact, extensive efforts were made to involve the public. Much to its credit, the voluntary process attracted numerous potential participants. It is impossible to know whether selection of another finalist community, keeping open options for numerous potential communities, or fuller commitment to extended public discussion might have led to a different outcome. Nonetheless, the compact's determination to proceed rapidly to a siting decision and US Ecology's adamancy about Butte in the face of the warning signs of conflict should not lead to rejection of the Nebraska case as typical of siting failure. At least in the stages leading up to site selection, the Nebraska case resembles the Ontario LLRW case and the successful hazardous waste sitings.

The Ontario LLRW siting case indicates that many of the lessons drawn from hazardous waste facility siting can be applied to the contentious issue of radioactive waste disposal. The American experience generally confirms the overall pattern of failed siting, although the Nebraska case suggests the possibility of implementing some version of the voluntary approach. Some states and compacts have begun to explore this approach, most notably Connecticut.[21] Whether these states will extend a general invitation for volunteers or attempt to develop the more comprehensive process necessary to enhance the prospects of extended dialogue over siting options and increase the likelihood of siting agreements remains unclear.

6

Toward a More Mature System

HAZARDOUS WASTE FACILITY SITING is often portrayed as a zero-sum game, in which a community that opens a waste treatment or disposal facility is the loser in an adversarial political process. Given the way most provinces, states, and waste management corporations have gone about the practice of siting, prevalence of the zero-sum view is entirely understandable. In most instances, siting decisions are made in the absence of prior community consultation and are not linked to other components of the hazardous waste management process. Instead, one community is singled out to assume much, or all, of the waste management burden for a number of wastes from a variety of regions, none of which may be well defined. A complicated form of redistributive politics results. When siting is pursued in this fashion in provinces or states, the prospects for reaching siting agreements are virtually nonexistent. Few communities will want to manage wastes generated by numerous industries and other communities, regardless of generous compensation packages that may be dangled before them by corporations or coercive powers of state or provincial authorities with which they may be threatened.

However, hazardous waste facility siting and management can be viewed as a positive-sum game. Hazardous waste is widely generated by industry, government, small business, and individual households in Canada and the United States. Approximately one ton of this waste is generated per person per year in each nation. Safe management is a collective responsibility, just as many of the material benefits derived from its generation are collective. This means that those organizations that generate the largest portion of the wastes must bear the largest portion of the costs of safe treatment and disposal. But the question of facility siting and the larger issue of waste management options must be placed in a broader political context, recognizing that all can win in a positive-sum game.

The cases in which positive-sum politics have prevailed is admittedly small, but the steady growth in their number, on both sides of the forty-ninth parallel, suggests that a transition may be within reach. Given the pessimistic tone of much of the literature on the issue of hazardous waste management, no success stories were expected to be found. Nonetheless, cases such as those in Alberta, Manitoba, Greensboro, and Minnesota suggest that traditional regulatory or market approaches need not be the only way to proceed. Furthermore, a growing body of evidence from Western Europe suggests striking similarities between the successful North American cases and the patterns of facility siting and waste management practiced abroad.[1] The Canadian and American siting successes are not fluke occurrences, and they point to a viable alternative route for future policy.

Perhaps one of the most remarkable aspects of hazardous waste policy is its relative newness in comparison with other spheres of domestic policy. Both Canada and the United States, for example, have been debating and exploring alternative methods for providing health insurance and pensions to senior citizens for much of the twentieth century. Even other areas of environmental policy, such as water and air pollution, have been on Canadian and American political agendas for decades, with well-established patterns and traditions for regulatory policy. By contrast, hazardous waste management and facility siting remain relative newcomers to the political agenda, as neither nation had any semblance of policy on these matters as recently as 1975. Subsequently, much has happened in a hurry, some of it useful but much of it wasteful in the extreme. Nonetheless, Canada and the United States, in their distinctive ways, could make significant strides toward a more effective set of hazardous waste policies in the decade ahead.

The shared design characteristics of a more mature approach emphasize the need for early and extensive public participation and involvement in a voluntary siting process; new institutional mechanisms for fostering dialogue over hazardous waste options; compensation and safety packages that make site acceptance more attractive and secure; and protections against exploitation, including assurances to any prospective site host communities that they alone will not bear the burden for waste management for an entire province, state, or region.

Canada: The Virtues and Pitfalls of Pure Decentralization

The history of hazardous waste management in Canada reflects the strengths and weaknesses of taking a decentralized approach to regulatory policy. For constitutional and political reasons, the federal government has continued to view many of the most important aspects of environmental regulation, including hazardous waste facility siting, as provincial responsibilities. It tends to intervene only in ways that will not conflict with ongoing provincial efforts, consistent with a general philosophy of "cooperative federalism." [2] As hypothesized by prevailing theories of intergovernmental relations, decentralization fosters considerable policy innovation and differentiation among separate units of the federal government. Under this approach, Canadian provinces become "laboratories of democracy," just as Justice Louis Brandeis envisioned the role of states in American federalism several generations ago.

The latitude given the provinces has conferred significant advantages, notably in the breakthrough siting agreements in Alberta and Manitoba. Both provinces serve as models for other provinces and states in developing genuine public debate over hazardous waste management options and securing broad public support for eventual siting and related waste management decisions. Whereas facility siting and waste management remain contentious issues in most other provinces and states, Alberta and Manitoba are resolving their management problems in a politically open and technologically sound manner. Moreover, the core components of their approach to siting and waste management have begun to diffuse. Even the Canadian federal government has carefully revised its historically contentious approach to low-level radioactive waste (LLRW) facility siting after the successful hazardous waste experiences of Alberta and Manitoba. Through this method, the government appears on the verge of a historic breakthrough in siting one or more disposal facilities for low-level radioactive wastes in Ontario. If successful, this would be the first new facility opened for either low-level or high-level radioactive wastes in either Canada or the United States in more than two decades.

A pair of successful provincial cases, however, does not mean a mature, nationwide system for hazardous waste management has been constructed. While in the United States a strong federal regulatory presence assures greater uniformity in waste management practices, the decentralized Canadian approach results in hazardous wastes being managed in a fashion that is not much safer than a generation ago. Lacking the infrastructure of landfills and incinerators that were developed in many parts of the United States before siting became controversial, a number of provinces have little or no access to sophisticated treatment or disposal capacity.

Several provinces that generate significant volumes of wastes have attempted siting, but with little success. The British Columbia and Ontario cases demonstrate that adherence to traditional market or regulatory siting practices appear fated to Nimby-type gridlock. Both provinces, as well as a number of smaller ones, must export their wastes, encourage on-site storage or disposal that is suspect and goes largely unregulated, or ignore the continuance of past disposal practices (including the placement of hazardous waste in sanitary sewers and solid waste landfills).

In the absence of federal pressure to assure proper facility operation or provincial commitment to develop a comprehensive system of waste management, Quebec's approach to hazardous waste appears little more refined in the 1990s than in the 1970s. Its major waste management facility has become a magnet for primarily American and non-Quebec Canadian wastes, because no system carefully monitors the wastes generated within the province and requires generators to secure appropriate treatment or disposal.

This mixed record of performance is also evident at the federal level. On the one hand, Ottawa has remained sufficiently distant from provincial practice to allow the Alberta and Manitoba approaches to be developed and implemented successfully. Moreover, the federal government has resisted the development of massive programs to clean up abandoned hazardous waste dumps, avoiding the American experience with Superfund. Cleanups have been left to the provinces, where Alberta and Manitoba have again led the way. And Ottawa has also learned from the provinces in developing a system of low-level radioactive waste facility siting that appears far superior in building public trust than anything attempted in the United States by individual states or multistate compacts.

On the other hand, the virtual absence of Canadian federal government involvement in hazardous waste policy has led to some serious problems. Environment Canada lacks the regulatory tools of the U.S. Environmental Protection Agency to prod laggard provinces to adhere to minimal standards of waste management and has done little basic data collection that is essential to understanding the full nature of the hazardous waste problem in Canada. Reliable measures of the volumes and types of waste generated vary greatly by province; comparable data in the United States are more dependable. Moreover, the Canadian government has been far more lax in playing a constructive role in promoting waste reduction and recycling than in the United States. The American federal role has not been stellar, but more incentives have been provided and greater pressures have been imposed on states than by the Canadian federal government.

The United States: The Mixed Record of Conjoint Federalism

The American approach to hazardous waste management reflects a greater sharing of powers between federal and state governments. Whereas the Canadian pattern of cooperative federalism indicates a gentle federal regulatory presence, American federalism is far more "conjoint," involving greater integration of responsibilities.[3] In hazardous waste facility siting, considerable decentralization continues to be provided, giving states virtually as much latitude as their provincial counterparts. But in other areas, such as cleanup of abandoned hazardous waste dumps, the federal government presence is far beyond anything attempted in Canada.

As with the Canadian approach, the American conjoint approach to regulatory federalism offers a mixed record. States have been left free to innovate in siting. While they have attempted a variety of initiatives, some complex and creative in nature, almost all have failed. The few recent successful American cases bear strong resemblance to the voluntary, comprehensive approach developed in Canada. When American states begin rethinking seriously how they go about the process of facility siting in the future, they might best turn to Alberta and Manitoba for guidance.

States have not been alone in their failure. The federal Superfund program is a drain on scarce resources and a magnet for litigation and adversarial politics. As currently structured, Superfund assigns responsibility for cleanup costs to those that can be identified as potentially responsible parties (PRPs). However, many PRPs face huge cleanup costs, often despite a relatively modest contribution to the total contamination problem. What commonly ensues is litigative finger pointing, involving respective PRPs, their insurers, and government agencies, not constructive dialogue over sharing cleanup responsibility in a timely fashion. The limited amount of site cleanup and the tremendous controversy over Superfund have caused enormous damage to the public's willingness to engage in any serious dialogue with public or private officials over waste management options, including developing a voluntary approach to siting.

The United States has made some strides toward a more effective hazardous waste management system since the early 1980s. Unlike Canada, the U.S. federal government and states are beginning to gain a clear picture of the volumes and types of hazardous waste that are being generated. Moreover, many of the most abusive waste disposal facilities have been closed, some having become Superfund sites. Similar facilities continue to operate in parts of Canada. For LLRW, some signs of interstate coordination have emerged, such as the creation of nine compacts involving forty-two states.

Perhaps most impressive are the significant strides that a number of states, corporations, and the federal government have made toward hazardous waste reduction and recycling. With the exceptions of Alberta and Manitoba, pollution prevention is far better defined and more aggressively pursued by federal and state programs in the United States than their federal and provincial counterparts in Canada. Both Ontario and Ottawa have rushed in 1993 and 1994 to catch up, but they remain far behind developments in many parts of the United States. The American system has not begun to tap the full potential of preventing waste; it instead has reacted to waste generation. But the case of Minnesota, and others, suggests that the conjoint federalism approach to hazardous waste has advanced waste reduction and recycling more rapidly than the cooperative federalism of Canada, with the exceptions of Alberta and Manitoba.

Policy Options for the 1990s and Beyond

The important differences between the Canadian and American experiences should not obscure the strong signs of convergence. A virtual common market in hazardous waste trading, technology, corporate involvement, and policy diffusion between states and provinces has emerged. As a result, a number of recommendations can be made for establishing future siting policy in both nations. The goal is to move beyond the adversarial politics that has dominated siting deliberations — and most aspects of hazardous waste management — in Canada and the United States. The new approach must be broadly inclusive, capable of generating public trust in a sphere of environmental policy that has become synonymous in many states and provinces with corruption and incompetence.

Begin a Dialogue

Perhaps the most difficult aspect of the proposed transition would entail the creation of early, open, and extensive provisions for public information and participation. In one sense, environmental policy is more open to public participation in the United States than in any other Western democracy because of its expansive definition of legal standing, provisions for open hearings, and public access to government documents. However, much of that involvement occurs after key decisions have been made. Often the process becomes a magnet for stalling tactics, including litigation, with the main public role confined to delay of program implementation, whether it involves siting or site cleanup.

The Canadian siting approach used so successfully in Alberta and Manitoba offers a model of participation that is designed to foster genuine democratic deliberation in examining waste management options in a variety of settings. Whereas most siting policies narrow deliberation immediately to a proposal for a single facility in a single community, these provinces successfully broadened the scope of discussion. In Alberta and Manitoba, final site selection involved fairly small communities, one of which is remote. But siting agreements need not be confined to isolated hamlets. Among the list of candidates that remained willing to explore the possibility of allowing a facility siting was the Manitoba capital, Winnipeg (population 560,000), suggesting that the process can accommodate a diverse set of citizens and communities. Subsequently, Winnipeg and another large Albertan city, Calgary, have accepted regional transfer facilities; yet another of these has been located just outside of Edmonton (population 574,000). Other successful facility siting efforts involved a city with nearly 200,000 residents (Greensboro) and a moderately sized suburb of a major city (Blainville), as well as smaller communities. Pending LLRW agreements in Ontario involve two small communities and two suburban-type communities with respective populations greater than 10,000 and 34,000. Under a process that emphasizes collective deliberation, agreements can be devised, in diverse settings, that address legitimate community concerns and long-term waste management needs.

Political scientists have correctly noted that past American and Canadian efforts to foster greater public deliberation and input into policymaking have often borne disappointing results. James A. Morone, for example, has lamented the recurring "democratic yearning" or "democratic wish" in America to make some strides toward opening up avenues of political participation, while failing to address fundamental questions of political power and equity in the process.[4] Further, Thomas E. Cronin and others have contended that recent increased use of direct democracy tools such as initiative and referendum in the United States falls prey to many of the same sorts of abuses in a representative democracy.[5] Canadian efforts to expand participatory opportunities have encountered comparable sorts of problems, compounded by the absence of formal participation procedures or traditions in many policy areas.[6]

Such cautionary statements, however, need not preclude serious exploration of more effective mechanisms to provide public participation and foster serious deliberation over complex and controversial public concerns, even ones as contentious as hazardous waste facility siting. The Minnesota, Greensboro, and California cases indicate that at least some aspects of the Alberta and Manitoba approach can be successfully transferred to the United States. In such instances, and even in the Nebraska low-level radioactive waste siting case, far-reaching efforts to involve and engage the public early in the facility siting process fostered a prolonged public dialogue absent in other American siting cases. And, at least in the instances of Minnesota and Greensboro, the outcomes — an emphasis on pollution prevention in the former and a new treatment facility in the latter — defied the typical Nimby pattern of gridlock and policy inertia.

Use Only Voluntary Siting

Extensive public education and participation are unlikely to result in siting agreements if they merely set the stage for top-down site selection by private corporations or governmental authorities. Successful siting agreements had voluntarism in common: the commitment by siting proponents that siting will proceed only in communities that demonstrate a clear willingness to accept facilities. In Swan Hills, Montcalm, and Greensboro, as well as communities in Ontario, assurances were made that siting proposals could be rejected at any time. In the Canadian cases, individual communities had to go on record as being open to the possibility of site acceptance before any further deliberations could proceed. The process led to harmonious agreements with individual communities and multiple volunteers of varying size and economic wherewithal. A voluntary process has considerable potential to develop a serious dialogue in an atmosphere of trust with multiple communities.

The emphasis on voluntarism went hand in hand with the public education and participation efforts in Alberta and Manitoba. Officials active in both siting efforts contend that these principles are mutually reinforcing, essential to developing basic trust. "The voluntary approach may not necessarily work everywhere, but I doubt siting can work at all if you try to just name the site," explained a senior Alberta official. "If anything is going to work, this is going to be it." A senior Manitoba official with comparable experience in the siting process in his province agreed: "Our voluntary approach was not a fluke. I'm not saying you can do it every time you try it, but I believe it can work in other places, in Canada, the United States, and elsewhere."

In many respects, the creation of a siting process that emphasizes dialogue and voluntarism offers a state and province time to explore fully its hazardous waste problems and its waste management options. No immediate treatment or disposal capacity crisis exists in certain regions of Canada or the United States; other areas have few good waste management choices. Thus rushing to try new treatment and disposal systems through market or regulatory mechanisms, which are unlikely to succeed in any event, is not necessary. The sort of process developed in Alberta and Manitoba encourages states and provinces to engage in serious examination of their overall waste management needs, instead of hurriedly exploring the desirability of a single facility that has been proposed for a single community. Through extensive deliberations, determinations can be made about the treatment and disposal facilities that are needed and the process that will be developed for educating the public and seeking volunteers, as in Alberta and Manitoba. Or it may be concluded after exhaustive public discussion that preventive waste reduction options offer the best alternative, as in Minnesota.

A voluntary approach is likely to succeed only if it is fully integrated with other components of a comprehensive waste management strategy. Sending letters to individual communities or placing advertisements soliciting volunteers, in the absence of extensive public participation measures and other key components of successful siting, remains unlikely to attract volunteers or secure siting agreements. States and provinces that take the easy approach to voluntarism may further compound the mistrust that is pervasive in most of the public discourse over hazardous waste facility siting and management. Such tactics probably will renew the common perception that facility siting is a zero-sum game, one that leaves the host a clear-cut loser, instead of part of a broader collective strategy that addresses waste management needs.

Require Burden Sharing

Traditional approaches to siting are further undermined by a common perception that fairness is lacking in the ways that burdens for waste management are distributed across a province, state, or region. Just as the general absence of public participation and a general emphasis on coercion instead of voluntarism contribute to Nimby reactions, a mounting perception that certain communities, states, and provinces bear a disproportionate share of the burden for waste management adds to the pervasive distrust of siting proposals and waste management strategies.

This perceived inequity takes two distinct forms. First, the distribution of treatment and disposal capacity tends to be concentrated in certain geographic regions. In the United States, large facilities are disproportionately located in select states, such as Alabama, Ohio, and South Carolina. The facilities have far greater capacity for treatment or disposal than industries, businesses, and households within their state boundaries generate, necessitating out-of-state waste import (often in substantial quantities) to maintain viable commercial operation. Other states, including many leading hazardous waste generators, have little if any such capacity within their own boundaries.

Most of the existing facilities were sited and opened in the 1960s and early 1970s, before hazardous waste became controversial. Many have become increasingly unpopular with local residents, as economic development and safety assurance packages were never included in the siting process or provided once facilities began operation. Ample evidence suggests that most residents in these communities had little if any idea of the sorts of activities that would take place within the facilities; they would be unlikely to support them if presented as new proposals today. Whereas communities such as Swan Hills and Montcalm continue to support their waste management facilities and feel that all aspects of siting were done in a fair, above-board manner, many communities that host the older sites have far less sanguine views about continued facility operation. These sentiments are increasingly evident in communities in both nations, which view themselves as having been unfairly targeted to shoulder substantial responsibility for waste management.

Second, a growing body of evidence indicates that waste management facilities, and the overall burden of waste management, tend to be concentrated disproportionately in those communities with large percentages of racial and ethnic minorities and residents with extremely low incomes and educational attainment levels.[7] As would be expected from North American land use practice, few hazardous waste incinerators or landfills are found on expensive sub-urban property. But increasingly sophisticated research suggests that, particularly in the United States, past siting efforts have not concentrated on areas in which broad awareness of and support for the siting existed or on ideal areas from a technical standpoint. Instead, it indicates that before siting became controversial, and communities of all racial and economic composi-tions became effective at using Nimby tactics, private waste management firms and government agencies targeted areas on the basis of their limited political and economic abilities to resist, not on important concerns such as maximizing safety. Consequently, a good deal of incineration and other forms of waste disposal take place in technically suspect places, including densely populated areas that are located near schools and other large concentrations of children. Much landfilling continues in areas where prospects for ground-water contamination remain high. More than 90 percent of hazardous waste is disposed of at or near the site of its generation, often in urban areas. Nearby residents appear unaware of the practice; different standards exist for evaluat-ing and approving of waste disposal; and the media and environmental advo-cacy groups have failed to draw attention to the issue.

The concerns over equity reinforce the common community reaction to market or regulatory siting proposals: anger and vigorous political response. Because siting is usually posed as a single proposition (some type of treatment or disposal facility) with few, if any, set boundaries concerning the types of wastes to be accepted or the geographic areas from which wastes will be re-ceived, communities are reluctant to volunteer themselves as magnets for massive quantities of wastes that other communities, states, provinces, or re-gions do not want. Under these circumstances, siting acceptance becomes a unilateral action of cooperation, with no commitment to waste management burden sharing with other communities. Time and again, siting opponents emphasized the unfairness of the situation. In most instances, they were given no prior notice that they were to be selected as a site host or told that siting would proceed only among volunteers, and no assurances were given that they would not become dumping grounds for vast quantities and types of wastes from other, sometimes far-removed, communities.

As in other aspects of hazardous waste management, the Alberta and Mani-toba cases provide models for surmounting burden-sharing concerns. Cana-dian federal law imposes no barriers on provinces to restrict waste exports or imports; nor does it stipulate any restrictions should multiple provinces choose to share responsibility for the wastes that they generate. Tight restric-tions on hazardous waste exports and imports contributed to the atmosphere of trust that developed during the Alberta siting process and were essential to win public support for siting in Swan Hills and other volunteer communities. Manitoba opted against a total prohibition, instead emphasizing preferential treatment for provincially generated wastes and providing a clause in the siting agreement to reopen the issue and explore the possibility of further restrictions should excessive imports become a problem. Manitoba also hopes to enter into a formal agreement with the Swan Hills facility to share technology and expertise.

Other provinces have not used the federal authority but remain flexible as they ponder future waste management strategies. In the United States, federal law is more restrictive. U.S. Supreme Court rulings would seem to foreclose the sorts of strategies devised by Alberta and Manitoba, although some constitutional uncertainty exists over whether state-funded and -operated facilities could impose such restrictions. Attempts in the late 1980s by states such as Alabama to either restrict waste imports or assess substantial tax surcharges on out-of-state wastes were deemed unconstitutional by decisive court majorities in 1992.[8] Given the rulings, incorporating Alberta- and Manitoba-type reassurances into the siting process becomes impossible for a state.

In theory, current American federal law provides some reassurance against waste management exploitation. States must demonstrate to the Environmental Protection Agency that they have secured adequate treatment and disposal capacity for a twenty-year period to receive federal Superfund dollars. But these capacity assurance reports are loosely structured documents, which often rely heavily on exportation and ambitious estimates of long-term waste reduction to achieve their assurance targets. Their findings often bear no reasonable relation to current within-state waste generation patterns, drawing into question the credibility of their long-term projections. Moreover, they have been gingerly enforced by an agency reluctant to cut off any flow of promised Superfund dollars and trigger a massive intergovernmental controversy.

Consequently, the federal government needs to examine burden sharing carefully as it begins the process of reauthorizing key pieces of hazardous waste legislation — the Resource Conservation and Recovery Act and the Superfund program. The Clinton administration has taken a first step through its February 1994 executive order that requires federal agencies to monitor the effects of their actions on environmental health in minority communities. A congressional change in the rules of interstate waste movement would be essential to establishing greater fairness in waste management and trust in future facility siting. Among the many options, two would seem particularly reasonable, not imposing tight federal clamps on interstate waste movement but promoting greater fairness and equity. First, Congress could levy a per mile fee on hazardous waste transport, from point of generation to point of disposal. The Supreme Court has already approved state policies that do this on an intrastate basis, and Congress could easily extend it nationally. A considerable disincentive would be provided against the highly undesirable practice of long-distance waste transport, and a supplemental source of revenue would be created for American hazardous waste programs. Costs would be imposed directly on waste generators unable to eliminate waste at the source or find a short-distance locale for final treatment or disposal.

Second, Congress might revisit creative policy proposals developed by a National Governors Association work group in the late 1980s. Going well beyond the capacity assurance provisions, each state would be encouraged to take serious steps toward developing effective systems of waste management instead of relying primarily on the easy option of export. The proposals called for the creation of "reasonableness criteria" that, in essence, deemed it "unreasonable for a state that needs to create capacity to fail to create that capacity for whatever reason." [9] Under these criteria, to continue to export wastes, a state must use out-of-state facilities only through formal interstate or regional agreements, make significant steps toward waste reduction through its policies, or construct and operate its own treatment and disposal facilities. [10]

The reasonableness criteria would leave enormous flexibility for individual state initiative or multistate partnership and would build on the most successful aspect of federal LLRW legislation: creation of interstate agreements through compacts. An array of choices would be available, including emphasizing waste reduction before considering siting, such as in Minnesota. But individual states would feel far more pressure than under current capacity assurance provisions to examine waste management options seriously and take greater responsibility for waste generated within their own boundaries. As to future siting efforts, if individual states decided that they needed to add treatment or disposal capacity, they could approach their citizens through a voluntary siting process that could provide assurance that any new site would not become a magnet for unreasonable — and uncontrollable — types and amounts of hazardous waste.

The option of regional approaches could lead to the construction of creative waste management partnerships, much like the one eminently possible among provinces such as Alberta and Manitoba. Regional agreements might be expanded to include one or more Canadian provinces as part of a larger burden-sharing arrangement. Canada and the United States in many respects already constitute an open, binational market of hazardous waste management. Shipping wastes from province to state (or vice versa) instead of within a nation may be more appropriate, given travel distances and complementary treatment and disposal capacity. Ontario and Quebec are the best candidates to participate in such an arrangement because of their already sizable role in this area. But waste trading should take place only under a formalized agreement, such as an extension of some form of the reasonableness criteria to participating Canadian provinces. This could create a regional partnership in which the waste management burden is shared fairly instead of deposited on a particular community.

Reduce Waste and Support Technological Innovation

Most of the siting proposals advanced under regulatory or market approaches involve use of traditional waste management technologies such as landfill, incineration, and deep-well injection. These remain the dominant technologies in current waste management practice, either on the site of generation or in commercial off-site facilities. An array of treatment technologies exist that may pose far less environmental or public health threats, yet they remain to be fully tested and utilized. Nimby rejection of all types of facilities and the enduring preference for traditional technologies by regulatory agencies, waste generators, and waste management corporations have served to delay the testing and development of the technological alternatives.

Any shift in hazardous waste policy must include systematic development of innovative technologies, beginning with efforts to maximize waste reduction. As demonstrated in the Minnesota case, a state (or province) has an array of incentives and related regulatory options that can be used to promote waste reduction in place of facility siting as a primary waste management strategy. Similarly, in the Alberta and Manitoba cases, waste reduction was integrated into the process of overall waste management and facility siting, and in other cases, major breakthroughs were seen in waste reduction that lessened the need for additional treatment or disposal capacity.

Individual firms, either in conjunction with public policy initiatives or independently, have discovered an unexpected ability to lower hazardous waste volumes and substitute materials to reduce the toxicity of the wastes that are generated. The New York–based environmental research group Inform and the congressional Office of Technology Assessment have played lead roles in examining the prospects for far-reaching reduction, much as the environmental group Pollution Probe has in Canada.[11] A 1993 Inform study found that manufacturing plants can reduce their hazardous waste output by 50 percent or more, often through fairly modest management or equipment alterations. Among numerous possible examples, a New Jersey–based Exxon plant cut its emissions of several toxic chemicals by 90 percent and saved $200,000 a year in avoided losses of raw materials through the simple action of placing covers over chemical holding tanks located around the plant. A major Ciba-Geigy plant was able to eliminate heavy-metal waste discharges by moving necessary purification steps to earlier stages of the process.[12] Significant waste reduction advances may be possible through expanded governmental, corporate, and community emphasis on this approach as a central component of a waste management strategy.

Even if the most optimistic estimates of waste reduction are realized, which seems unlikely in the near term, future policymakers must consider alternative treatment and disposal technologies for managing the remaining wastes. An array of new technologies are being used in Western Europe but remain largely on the fringe of North American waste management. Biological, chemical, neutralization, and physical treatment processes have been designed to reduce or eliminate the hazardous properties of any materials that remain after treatment. Thermal technologies, for example, stop short of burning waste, breaking down chemical compounds into inert elements instead of creating the hazardous by-products that are the result of most forms of incineration. Further options include development of household and community waste pickup programs, regional transfer and storage facilities (as developed in Alberta and Manitoba), experimental solidification and stabilization facilities (as developed in Quebec), transportable treatment and disposal units, and advanced landfills such as residuals repositories.[13]

State and provincial efforts to inform and involve the public in waste management planning should explore alternative technologies and not focus exclusively on a major facility proposal offering a single, traditional technology. Future waste management could involve a blending of waste reduction efforts with creative treatment and disposal technologies that minimize environmental damage and public health risk and that contain costs. Such a strategy offers a broader distribution of waste management across a geographic area, involving all generators of waste management and a variety of innovative treatment and disposal facilities and thus enhancing the prospects for burden sharing and assurances to host communities that they will not be exploited.

Raise Institutional Credibility

Most public agencies and private firms responsible for waste management and facility siting in Canada and the United States suffer from low public credibility. Their traditional approach to facility siting has only exacerbated earlier distrust. Both Alberta and Manitoba have eliminated the credibility problem by creating crown corporations and recruiting policy professionals with considerable skills of risk communication. Other provinces or states may not need to follow the same strategy or play a direct and enduring role in the management of any facility, but they must seek new institutional mechanisms that can formulate and implement waste management policy in an atmosphere of greater trust. In the United States, state-based public utility commissions may serve as a viable institutional model.[14] They oversee electric power generation, distribution, pricing, and conservation, although they do not assume the fuller responsibilities of Canadian crown corporations.

Emphasizing public participation and voluntarism in siting could enhance the credibility of the institutions and professionals responsible for guiding hazardous waste policy. Publicly negotiated legal agreements between a community hosting a facility and the corporation or agency (or both) that will be responsible for facility operation, such as in the Alberta and Manitoba cases, also promote institutional credibility. Under a comanagement agreement or a performance contract, the site proponent spells out all terms pertinent to long-term facility management, including compensation and safety guarantees, provisions for ongoing public participation, the types of technologies to be used at the facility and wastes to be accepted, and provisions for facility closure and protection. As in the Manitoba case, many clauses are drafted to allow the community to reexamine key aspects of the agreement in the future should concerns arise, such as excessive out-of-province imports. Such an agreement empowers a host community in a way that traditional siting arrangements do not. The role of public participation becomes an enduring one, not confined to an up-or-down vote on a specific proposal. If private waste management firms and public agencies are sincere about their claims of growing sophistication and reliability in the utilization of waste management technology, they should be comfortable with drawing up legal agreements. Formal arrangements and a willingness to allow prolonged discussion with communities would provide firms and agencies with an opportunity to demonstrate their commitment to a genuine waste management partnership and their desire to divorce from the adversarial relationships of the past.

Trash Superfund

Superfund may be the quintessential example of flawed environmental policy design. Drafted initially in the waning days of the Carter administration as a temporary, $5 billion cleanup for abandoned hazardous waste dumps, the program has evolved into a litigative search-and-destroy mission that gobbles large quantities of federal, state, and private dollars that could be put toward other, more purposeful, waste management uses. Superfund fuels the distrust and acrimony that have made any prolonged dialogue over siting and waste management options difficult. A 1986 reauthorization made matters worse, intensifying the earlier emphasis on technically unattainable standards and litigative battles to pinpoint responsibility for waste management transgressions of prior generations.[15]

After more than a decade of implementation and expenditures of billions of private and public dollars, Superfund has cleaned up less than 5 percent of the most polluted abandoned dumps in the United States. The pace has increased in recent years, but numerous studies contend that vast percentages of Superfund dollars go to litigative costs and consulting contracts instead of treatment and disposal of wastes at these sites. For example, a 1992 Rand Institute study concluded that nearly 90 percent of the money spent by insurers on Superfund-related claims has been devoted to legal and related costs.[16]

Canada has largely dodged the issue of abandoned dump cleanup, both for better and for worse. No federal legislation has been enacted; the matter has been left to provinces. With few exceptions, decisions have yet to be made about the waste dumps. The Alberta experience may offer some insight into future policy. Alberta has tried to avoid the counterproductive finger pointing prevalent in Superfund. At about the same time the province's central waste management facility was being opened, Alberta launched Help End Landfill Pollution (HELP). The program involved a four-phase process, beginning with an inventory of abandoned industrial landfill sites, many of which were contaminated with large quantities of hazardous waste. After an in-depth assessment, preventive action and cleanup were pursued, in conjunction with an effort to place the financial burden of cleanup on the landfill owner, where identifiable. Many of these wastes have ultimately been shipped to Swan Hills for treatment or disposal. This approach has led to far more extensive and rapid site cleanup than anything experienced thus far in the United States or the rest of Canada.

Looking ahead, the United States needs to consider fundamental reform of the Superfund program, while Canada and many of its provinces must begin to wrestle with abandoned wastes. The Alberta example suggests some promise for a focused cleanup effort, linked to the availability of a central facility, regional storage areas, and a unified system of waste transport. In the United States, thoughtful proposals have begun to circulate that incorporate some of the shared design characteristics crucial to siting agreements, including an emphasis on public involvement by giving individual communities earlier and more direct input into the types of cleanup efforts they prefer. In addition, some proposals explore alternative burden-sharing arrangements by proposing more equitable methods for allocating cleanup costs.[17]

Without meaningful reform of Superfund, the emergence of a more mature set of waste management policies in the United States, much less the beginning of serious dialogue over management options, is difficult to envision. Perhaps the greatest single difference between Canada and American hazardous waste policy is that the United States has Superfund, which consistently erodes available funds and remaining public trust. Canada, meanwhile, remains largely free to chart its own course, and, as in other areas of environmental policy, learn from prior American mistakes.

Environmental Regulatory Integration

Hazardous waste policy has been viewed in Canada and the United States as a great drain on resources and energies that could be concentrated on more pressing environmental problems. Repeated surveys of environmental experts point to the high levels of resources and public attention directed toward hazardous wastes, even though numerous other environmental concerns could pose greater public health problems. Nonetheless, hazardous waste management probably will continue to consume a growing share of the resources that federal, state, and provincial governments, and thousands of private firms, devote to environmental protection.

Much current activity, public and private in both nations, is expensive but of minimal environmental or public health benefit. Since the mid-1970s public distrust has become so great and public policy so flawed that having a constructive discussion about waste management options and the design and implementation of an efficient and effective policy is nearly impossible. Canada and the United States have designed similar approaches to siting and have experienced similar, Nimby-type results. Alternatives do exist that deviate from prevailing approaches to siting and waste management and, in many cases, deliver different outcomes. While no easy answers are available for the complex problems involved with siting and waste management, the future of hazardous waste management need not be characterized by political, legal, and social wranglings.

A new, more mature era of waste management could involve participatory, institutional, and technological changes and provide an opportunity to begin to integrate hazardous waste policy with other spheres of environmental policy. Every administrator of the Environmental Protection Agency, from William D. Ruckelshaus in the Nixon administration to Carol M. Browner in the Clinton administration, has lamented the fragmented nature of environmental regulatory legislation and the resulting bureaucratic structure. Similar woes have been heard from their Canadian federal counterparts, as well as from those at the state and provincial levels.

By any measure, environmental regulatory policy in both nations is highly fragmented, with the most clear-cut divisions being made among the mediums of air, water, and land. Pollutants and wastes defy medium-based boundaries, but U.S. and Canadian policies (at federal, state, and provincial levels) have largely adhered to the artificial categorizations. Consequently, separate programs with separate agency subunits and related interest groups emerge. Upon implementation of the programs, a shift of pollutants or toxic substances occurs from one medium to another. For example, certain air pollution control technologies improve air quality but, in the process, generate a concentrated by-product that is a highly toxic form of hazardous waste. Through treatment or disposal, the waste shifts to another medium, such as into the air through incineration or onto land and groundwater through landfill. For many American and Canadian industries, a majority of the hazardous waste that they generate is created as part of their compliance process with medium-specific environmental regulations. Thus, instead of addressing environmental problems in an integrated manner, this fragmented system results in pollutants and toxic substances merely being shuffled to the least regulated medium at a given moment.[18]

One of the more heartening developments in both Canada and the United States in recent years has been the growing recognition of the multimedia quality of environmental policy and a growing set of initiatives designed to better integrate the numerous stages of the regulatory process. Hazardous waste management has generally been excluded from these efforts, given its political volatility. Nonetheless, a more comprehensive system of hazardous waste management could go hand in hand with a more integrated system of environmental regulation, offering extraordinary environmental, public health, and economic benefits.

Looking beyond Waste

The acronym *Nimby*, and all its attendant political causes and consequences, cannot be confined to hazardous waste facility siting, or even environmental policy generally. The tendency for communities to act decisively and halt proposed developments is an increasingly common feature of the American and Canadian political experiences, crossing an array of policy areas. In the case of hazardous waste, the Nimby reaction is reasonable and understandable, given the way proposals are handed down in the absence of prior consultation from either governmental authorities or private sector entrepreneurs. But as the Nimby response becomes increasingly common, a larger question is raised of how nations such as Canada and the United States might better go about allocating responsibility for siting facilities that provide important public services.

The Nimby phenomenon is found in an variety of settings, including nursing homes, hospices for the terminally ill, drug and alcohol treatment facilities, recreational facilities, prisons, solid waste recycling facilities, housing designed primarily for low-income citizens or allowing children as residents, and churches. The politics underlying specific proposals differ from case to case, but the general reliance on top-down, regulatory or market approaches appears to prevail. Communities are given little advance warning of these proposals and little or no opportunity to engage in extended dialogue over their desirability, conditions that would be attached to their acceptance, or ways multiple communities could equitably share the responsibility of accepting the facilities and offering needed services.

Facility proponents will always be tempted to maintain a closed site selection process, assuming that either strong-arm regulatory tactics or economic compensation offers will silence any public concern once the selection is made. But the more open the process, the more likely the collective action problem is to be resolved in a cooperative fashion. The emphasis on collective deliberation enhances the likelihood of extended discussion and a serious search for solutions.

Within the process, governmental and private institutions can play essential roles. They can set up the process for deliberation and draw on their expertise to provide vital information. From the Alberta and Manitoba crown corporations to the Greensboro task forces, these institutions replaced traditional top-down relationships with the citizenry in favor of a partnership. As political scientist V. O. Key noted more than three decades ago, "Those leaders who act as if they thought the people to be fools responsive only to the meanest appeals deserve only scorn." [19] The experience of hazardous waste facility siting suggests that such leaders continue to play a dominant role in many provinces, states, and waste management corporations, but that other leaders — and institutional mechanisms — are available that can treat the public with respect and allow people to play a decisive role in resolving difficult policy problems. This approach may prove equally fruitful if applied to other policy areas.

Opening up a traditionally closed process is an essential step toward the sorts of collective deliberation necessary to move beyond the Nimby syndrome. But achievement of a more equitable, publicly acceptable distribution of social and economic costs must also entail development of burden-sharing arrangements. Instead of depositing controversial facilities in settings least politically and economically able to oppose them, siting must be democratically directed, determined on a voluntary basis, and accompanied by firm assurances that overall responsibility will be distributed in an equitable manner. No longer can problems generated in one community, state, or province be exported somewhere else. Protections must be provided against exploitation, such as calling for long-term community input into facility management and assurances that facilities will be safely operated and maintained.

Such a combination of features accentuates both the rights that citizens have to guide policymaking that affects their communities and their responsibilities to contribute to meaningful solutions. Unfortunately, while the dominant approaches to many areas of domestic policy have given communities greater reason to say no to proposed solutions that might impose some direct cost, they have not provided communities with any encouragement — or reason — to devise constructive solutions to pressing problems. As a result, "rights talk" trumps discussion of collective responsibility time and again. At its worst, according to legal scholar Mary Ann Glendon, the imbalances "incline us to shift the costs and risks of current policies regarding natural resources, pollution, public indebtedness, social security, and public health onto our children and future generations." [20] Costs and risks may also be shifted geographically, to other communities and regions.

Progress will not be easy, especially given the pervasive cynicism directed toward public institutions and policy professionals in both Canada and the United States. But available evidence indicates that transformation of the dominant approaches to siting can result in a change in the siting process. As one participant in the alternative siting process pioneered in Alberta noted, "Although the process started out with a focus on something undesirable ('waste'), the process addressed a need and resulted in the development of a feeling of belonging and connection in the community. Responsibility, pride and a positive attitude result in a stronger spirit of community and offer an intangible incentive for participation in the decision process." [21] This sort of political activity has not often been found in Canada or the United States recently. Ways are needed to replicate it if more mature methods of governance, in environmental policy and other spheres of domestic policy, are to be devised in the years ahead.

Notes

Chapter 1

1. R. Kent Weaver, "The Politics of Blame Avoidance," *Journal of Public Policy*, vol. 6 (October–December 1986), pp. 371–98.

2. Robert P. Stoker, *Reluctant Partners: Implementing Federal Policy* (University of Pittsburgh Press, 1991), p. 152.

3. "1992 Outlook for Commercial Hazardous Waste Management Facilities: A Nationwide Perspective," *Hazardous Waste Consultant* (March–April 1992), pp. 4.1–4.17; "1993 Outlook for Commercial Hazardous Waste Management Facilities: A Nationwide Perspective," *Hazardous Waste Consultant* (March–April 1993), pp. 4.1–4.21; and "1994 Outlook for Commercial Hazardous Waste Management Facilities: A North American Perspective," *Hazardous Waste Consultant* (March–April 1994), pp. 4.1–4.20.

4. Barry G. Rabe and John Martin Gillroy, "Intrinsic Value and Public Policy Choice: The Alberta Case," in John Martin Gillroy, ed., *Environmental Risk, Environmental Values, and Political Choices: Beyond Efficiency Tradeoffs in Public Policy Analysis* (Boulder, Colo.: Westview, 1993), pp. 150–70. I am greatly indebted to John Gillroy for his insights into this area.

5. Bruce A. Williams and Albert R. Matheny, *Democracy, Dialogue, and Social Regulation* (Yale University Press, forthcoming); and John S. Dryzek, *Discursive Democracy: Politics, Policy, and Political Science* (Cambridge University Press, 1990).

6. Paul E. Peterson, Barry G. Rabe, and Kenneth K. Wong, *When Federalism Works* (Brookings, 1986).

7. Paul J. Quirk, "The Cooperative Resolution of Policy Conflict," *American Political Science Review*, vol. 83 (September 1989), p. 908.

8. See, for example, Christopher J. Bosso, *Pesticides and Politics: The Life Cycle of a Public Issue* (University of Pittsburgh Press, 1987); Charles O. Jones, *Clean Air: The Policies and Politics of Pollution Control* (University of Pittsburgh Press, 1975); and R. Shep Melnick, *Regulation and the Courts: The Case of the Clean Air Act* (Brookings, 1983).

9. Elinor Ostrom, *Governing the Commons: The Evolution of Institutions for Collective Action* (Cambridge University Press, 1990), p. 14.

10. For a discussion of this challenge in the area of federal program implementation, see Stoker, *Reluctant Partners*, chap. 2.

11. James G. March and Johan P. Olsen, *Rediscovering Institutions: The Organizational Basis of Politics* (Free Press, 1989), p. 52.

12. This overlaps considerably with Ostrom's findings on the emergence of cooperation in management of common pool resources. Ostrom, *Governing the Commons*, chap. 6.

13. For a thoughtful review of these perspectives and their limitations, see Anne M. Khademian, *The SEC and Capital Market Regulation: The Politics of Expertise* (University of Pittsburgh Press, 1992).

14. On the political and policy ramifications of issue areas with high salience and high conflict, see David E. Price, *Policymaking in Congressional Committees: The Impact of "Environmental" Factors* (University of Arizona Press, 1979). See also Frank R. Baumgartner and Bryan D. Jones, *Agendas and Instability in American Politics* (University of Chicago Press, 1993).

15. Mary Ann Glendon, *Rights Talk: The Impoverishment of Political Discourse* (Free Press, 1991), p. 172.

16. Gail Bingham, *Resolving Environmental Disputes: A Decade of Experience* (Washington: Conservation Foundation, 1986); and Lawrence Susskind and Jeffrey Cruikshank, *Breaking the Impasse: Consensual Approaches to Resolving Public Disputes* (Basic Books, 1987).

17. Andrew S. McFarland, *Cooperative Pluralism: The National Coal Policy Experiment* (University of Kansas Press, 1993).

18. 90 Stat. 2799.

19. Richard C. Fortuna and David J. Lennett, *Hazardous Waste Generation: The New Era* (McGraw-Hill, 1987), pp. 26–27. For further discussion, see Christopher Harris, William L. Want, and Morris A. Ward, *Hazardous Waste: Confronting the Challenge* (New York: Quorum, 1987).

20. Linda Greenhouse, "Justices Decide Incinerator Ash Is Toxic Waste," *New York Times*, May 3, 1994, p. A1.

21. 1993 Minnesota Waste Management Act Amendments (Minn. Stat. 115A.916 [1993]); and "New Waste Laws Stress Source Reduction," *Resource: Perspective on Minnesota Waste Issues*, vol. 3 (July–August 1993), p. 9.

22. Transportation of Dangerous Goods Act (S.C. 1980–81–82–83, c. 36, as amended). For further discussion, see William M. Glenn, Deborah Orchard, and Thia M. Sterling, *Hazardous Waste Management Handbook*, 5th ed. (Don Mills, Ontario: Southam, 1988), pp. 26–27.

23. The council consists of the environmental ministers of each of the ten provinces, two territories, and federal government. Its primary contribution in hazardous

waste management has been to draw greater attention to the issue and the substantial variation among provincial approaches. The Canadian Environmental Protection Act (S.C. 1988, c. 22) provides authority to manage toxic substances from "cradle to grave," although in practice it continues to face "numerous hurdles to speedy or unilateral action." David Vanderzwaag and Linda Duncan, "Canada and Environmental Protection: Confident Political Faces, Uncertain Legal Hands," in Robert Boardman, ed., *Canadian Environmental Policy: Ecosystems, Politics, and Process* (Oxford University Press, 1992), pp. 9–10. For further discussion, see Alastair R. Lucas, "The New Environmental Law," in Ronald L. Watts and Douglas M. Brown, eds., *Canada: The State of the Federation 1989* (Kingston, Ontario: Institute of Intergovernmental Relations, 1989), pp. 167–92.

24. George Hoberg, "Sleeping with an Elephant: The American Influence on Canadian Environmental Regulation," *Journal of Public Policy*, vol. 11, no. 1 (1991), pp. 107–32; and Kathryn Harrison and George Hoberg, "Setting the Environmental Agenda in Canada and the United States: The Cases of Dioxin and Radon," *Canadian Journal of Political Science*, vol. 24 (March 1991), pp. 3–27. On the diffusion of trade and other economic policy ideas across state and provincial boundaries, see Douglas M. Brown and Earl H. Fry, eds., *States and Provinces in the International Economy* (Berkeley, Calif.: Institute of Governmental Studies, 1993).

25. For added discussion of the American approach to regulatory federalism, see U.S. Advisory Commission on Intergovernmental Relations, *Regulatory Federalism: Policy, Process, Impact, and Reform* (Washington, 1984).

26. Grace Skogstad and Paul Kopas, "Environmental Policy in a Federal System: Ottawa and the Provinces," in Boardman, ed., *Canadian Environmental Policy*, pp. 52–53. More generally on the Americanization of Canadian judicial decisionmaking, see Christopher P. Manfredi, *Judicial Power and the Charter: Canada and the Paradox of Liberal Constitutionalism* (University of Oklahoma Press, 1993).

27. David M. Welborn, "Conjoint Federalism and Environmental Regulation in the United States," *Publius: The Journal of Federalism*, vol. 18 (Winter 1988), pp. 28–43.

28. Thomas W. Church and Robert T. Nakamura, *Cleaning Up the Mess: Implementation Strategies in Superfund* (Brookings, 1993).

29. Daniel A. Mazmanian and David L. Morell, *Beyond Superfailure: America's Toxics Policy for the 1990s* (Boulder, Colo.: Westview, 1992), chap. 4.

30. Quoted in Charles Piller, *The Fail-Safe Society: Community Defiance and the End of American Technological Optimism* (Basic Books, 1992), p. 4. For a review of differing interpretations of the Nimby syndrome, see William R. Freudenburg and Susan K. Pastor, "Nimbys and Lulus: Stalking the Syndromes," *Journal of Social Issues*, vol. 48, no. 4 (1992), pp. 39–61.

31. Quoted in Mark David Richards, *Siting Industrial Facilities: Lessons from the Social Science Literature* (Washington: U.S. Council for Energy Awareness, 1992), p. 31. See also Michael O'Hare and Debra Sanderson, "Facility Siting and Compensation: Lessons from the Massachusetts Experience," *Journal of Policy Analysis and Management*, vol. 12, no. 2 (1993), pp. 364–76.

32. U.S. Environmental Protection Agency, *Everybody's Problem: Hazardous Waste* (Washington, 1980).

33. Barry G. Rabe, "Environmental Regulation in New Jersey: Innovations and Limitations," *Publius: The Journal of Federalism*, vol. 21 (Winter 1991), pp. 83–103.

34. U.S. Council on Environmental Quality, *Environmental Quality: The Twentieth Annual Report of the Council on Environmental Quality* (Washington, 1990), pp. 248–49; and Government Institutes, *Case Studies in Waste Minimization* (Rockville, Md., 1991).

35. Mazmanian and Morell, *Beyond Superfailure*, pp. 136–42; and Bruce W. Piasecki and Gary A. Davis, *America's Future in Toxic Waste Management: Lessons from Europe* (New York: Quorum, 1987).

36. "1992 Outlook for Commercial Hazardous Waste Management Facilities," pp. 4.1–4.47; and "1993 Outlook for Commercial Hazardous Waste Management Facilities."

37. Andrew Szasz, "Corporations, Organized Crime, and the Disposal of Hazardous Waste: An Examination of the Making of a Criminogenic Regulatory Structure," *Criminology*, vol. 24, no. 1 (1986), pp. 1–27; Alan A. Block and Frank R. Scarpitti, *Poisoning for Profit: The Mafia and Toxic Waste in America* (William Morrow, 1985); U.S. General Accounting Office, *Illegal Disposal of Hazardous Waste: Difficult to Detect or Deter*, Comptroller General's Report to the Subcommittee on Investigations and Oversight, Committee on Public Works and Transportation, U.S. House of Representatives (1985). Similar problems have emerged in solid waste management. See Jeff Bailey, "To Keep Prices High, Some Trash Haulers May Break the Rules," *Wall Street Journal*, November 8, 1993, p. A6.

38. Mazmanian and Morell, *Beyond Superfailure*, pp. 14, 106–07, 182–83; Jeff Bailey, "Concerns Mount over Operating Methods of Plants That Incinerate Toxic Waste," *Wall Street Journal*, March 20, 1992, p. B1; and Jeff Bailey, "Incinerators Take On Kilns in Hazardous Waste Battle," *Wall Street Journal*, November 22, 1993, p. B4.

39. Mazmanian and Morell, *Beyond Superfailure*, pp. 106–07.

40. William Gruber, "Solving the Equity Problem," *EI Digest: Industrial and Hazardous Waste Management* (July 1991), pp. 3, 6. See also Jonathan Walters, "The Poisonous War over Hazardous Waste," *Governing*, vol. 5 (November 1991), pp. 32–35.

41. Environmental Information, *Interdependence in the Management of Hazardous Waste* (St. Paul, Minn., 1993), p. 4.

42. Mazmanian and Morell, *Beyond Superfailure*, p. 86.

43. Timothy Noah, "EPA Unveils Plans to Curb Incinerators of Hazardous Waste by Blocking Growth," *Wall Street Journal*, May 19, 1993, p. B6.

44. Environmental Information, *Interdependence in the Management of Hazardous Waste*, section IV.

45. Kirk Victor, "Trouble on Wheels," *National Journal*, May 26, 1990, p. 1288.

46. "Canada Cracking Down on Hazwaste Haulers," *Great Lakes Reporter*, vol. 7 (September–October 1990), p. 10; and "Hitting the Road: Hazardous Waste Hauling on the Rise," *Great Lakes Reporter*, vol. 6 (March–April 1989), pp. 1, 3.

47. John Jackson and Phil Weller, *Chemical Nightmare* (Toronto, Ontario: Between the Lines, 1982), p. 19.

48. Doug Macdonald, *The Politics of Pollution: Why Canadians Are Failing Their Environment* (Toronto, Ontario: McClelland and Stewart, 1991), pp. 216–18.

49. Richard N. L. Andrews, Raymond J. Burby, and Alvis G. Turner, *Hazardous*

Materials in North Carolina: A Guide for Decisionmakers in Local Government (University of North Carolina Institute for Environmental Studies, 1985), pp. 130–31.

50. Mary Tiemann, *Waste Exports: U.S. and International Efforts to Control Transboundary Movement* (Washington: Congressional Research Service, 1989), p. 4.

51. For an overview of waste exports on the continent, see Barry G. Rabe, "Exporting Hazardous Waste in North America," *International Environmental Affairs: A Journal for Research and Policy*, vol. 3 (Spring 1991), pp. 108–23.

52. Jeffrey D. Smith, "The Ontario Industrial Waste Market," *EI Digest: Industrial and Hazardous Waste Management* (May 1994), p. 15.

53. William Gruber, "North Carolina's Hazardous Waste: Pursuing a 'We Make It, You Take It' Policy," *EI Digest: Industrial and Hazardous Waste Management* (September 1991), pp. 16–21.

54. Robert D. Bullard, *Dumping in Dixie: Race, Class, and Environmental Quality* (Boulder, Colo.: Westview, 1990), p. 9.

55. Ibid., p. 108.

56. Michael R. Greenberg, Richard F. Anderson, and Kirk Rosenberger, "Social and Economic Effects of Hazardous Waste Management Sites," *Hazardous Waste*, vol. 1, no. 3 (1984), p. 387. For more recent confirmation of this pattern, see James T. Hamilton, "Politics of Social Costs: Estimating the Impact of Collective Action on Hazardous Waste Facilities," *Rand Journal of Economics*, vol. 24 (Spring 1993), pp. 101–25; and Bunyan Bryant and Paul Mohai, eds., *Race and the Incidence of Environmental Hazards* (Boulder, Colo.: Westview, 1992).

57. Diane Austin, "Knowledge and Values in the Decisionmaking around Hazardous Waste Facilities," University of Michigan School of Natural Resources and Environment (1991); and Robert Tomsho, "Indian Tribes Contend with Some of Worst of America's Pollution," *Wall Street Journal*, November 29, 1990, p. A1.

58. See, for example, David Vogel, *National Styles of Regulation: Environmental Policy in Great Britain and the United States* (Cornell University Press, 1986); Ronald Brickman, Sheila Jasanoff, and Thomas L. Ilgen, *Controlling Chemicals: The Politics of Regulation in Europe and the United States* (Cornell University Press, 1985); and David Vogel, "Representing Diffuse Interests in Environmental Policymaking," in R. Kent Weaver and Bert A. Rockman, eds., *Do Institutions Matter? Government Capabilities in the United States and Abroad* (Brookings, 1993), pp. 237–71.

59. See, for example, Hoberg, "Sleeping with an Elephant," pp. 107–32; and Harrison and Hoberg, "Setting the Environmental Agenda in Canada and the United States," pp. 3–27.

60. Seymour Martin Lipset, *Continental Divide: The Values and Institutions of the United States and Canada* (Washington: Canadian-American Committee of C. D. Howe Institute and National Planning Association, 1989); Richard M. Merelman, *Partial Visions: Culture and Politics in Britain, Canada, and the United States* (University of Wisconsin Press, 1991); and Christopher Leman, *The Collapse of Welfare Reform: Political Institutions, Policy, and the Poor in Canada and the United States* (MIT Press, 1980).

61. Lipset, *Continental Divide*, p. 3.

62. Lucas, "The New Environmental Law"; and Mary Williams Walsh, "Environmental Law in Canada Comes of Age," *Los Angeles Times*, April 8, 1990, p. D1.

63. Boardman, ed., *Canadian Environmental Policy.*

64. Richard Johnston, *Public Opinion and Public Policy in Canada: Questions of Confidence* (University of Toronto Press, 1988); and Robert Burge, "A Question of Confidence: Revisited," paper prepared for the 1993 annual meeting of the Canadian Political Science Association.

65. David A. Olsen and C. E. S. Franks, *Representation and Policy Formation in Federal Systems* (Berkeley, Calif.: Institute of Governmental Studies, 1993).

66. Benjamin Ginsberg and Martin Shefter, *Politics by Other Means: The Declining Importance of Elections in America* (Basic Books, 1990); and Morris Fiorina, *Divided Government* (Macmillan, 1992).

67. Donald V. Smiley, "Public Sector Politics, Modernization, and Federalism: The Canadian and American Experiences," *Publius: The Journal of Federalism*, vol. 14 (Winter 1984), pp. 39–59; Ronald L. Watts, "Divergence and Convergence: Canadian and U.S. Federalism," in Harry N. Scheiber, ed., *Perspectives on Federalism: Papers from the First Berkeley Seminar on Federalism* (Berkeley, Calif.: Institute of Governmental Studies, 1987), pp. 180–208; and Milton J. Esman, "Federalism and Modernization: Canada and the United States," *Publius: The Journal of Federalism*, vol. 14 (Winter 1984), pp. 21–38.

68. Esman, "Federalism and Modernization," p. 29.

69. This finding is similar to that of David Mayhew's in many areas of domestic policy at the federal level. See David Mayhew, *Divided We Govern: Party Control, Lawmaking, and Investigations, 1946–1990* (Yale University Press, 1991).

70. Barry G. Rabe, "Cross-Media Regulatory Integration: The Case of Canada," *American Review of Canadian Studies*, vol. 19 (Autumn 1989), pp. 261–73.

71. Mario Ristoratore, "Siting Toxic Waste Disposal Facilities in Canada and the United States: Problems and Prospects," *Policy Studies Journal*, vol. 14 (September 1985), p. 141.

72. Mario Ristoratore, "Siting Toxic Waste Disposal Facilities: A Comparative Study of Public Policy in the United States and Canada," Ph.D. dissertation, Brandeis University, 1986, p. 237.

73. On this pattern of environmental regulatory diffusion, see Barry G. Rabe and Janet B. Zimmerman, *Toward Environmental Regulatory Integration in the Great Lakes Basin* (University of Michigan Resource for Public Health Policy, 1992).

74. Peter Moon, "Volume of Medical Waste Grows as Options for its Disposal Shrink," *Toronto Globe and Mail*, June 14, 1990, pp. A1, A4.

75. Andrew Porterfield and Jock Ferguson, "U.S. Bill Could End Waste Flow to Canada," *Toronto Globe and Mail*, August 19, 1989, p. A11.

76. Rabe, "Exporting Hazardous Waste in North America."

77. Fred Thompson, "Introduction," in Fred Thompson, ed., *Regulatory Regimes in Conflict: Problems of Regulation in a Continental Perspective* (Lanham, Md.: University Press of America, 1984), p. viii.

78. Thomas L. Ilgen, "Between Europe and America: Regulating Toxic Substances in Canada," in Thompson, ed., *Regulatory Regimes in Conflict*, pp. 1–30. For a slightly different version, see Thomas L. Ilgen, "Between Europe and America, Ottawa and the Provinces: Regulating Toxic Substances in Canada," *Canadian Public Policy*, vol. 11, no. 3 (1985), pp. 578–90.

79. Douglas M. Brown and Earl H. Fry, eds., *States and Provinces in the International Economy* (Berkeley, Calif.: Institute of Governmental Studies, 1993).

Chapter 2

1. For an early but enduring assessment of the limitations of regulatory approaches, see David L. Morell and Christopher Magorian, *Siting Hazardous Waste Facilities: Local Opposition and the Myth of Preemption* (Cambridge, Mass.: Ballinger, 1982).

2. This roughly parallels the process that so often leads to conflictual outcomes in game theoretical settings. One of the most widely used versions of game theory, the so-called prisoner's dilemma, recognizes the interdependency of political action and pinpoints many of the impediments to cooperative decisionmaking. It operates with only two distinct parties and no central authority, emphasizing the differential payoffs to participants that lead to highly desirable and understandable outcomes for them as well as the strong incentives that participants face to defect (or not cooperate). On game theory and its lessons for cooperation, see Robert Axelrod, *The Evolution of Cooperation* (Basic Books, 1984). On applications of game theory to hazardous waste policy, see Barry G. Rabe, "The Hazardous Waste Dilemma and the Hazards of Institutionalizing Negotiation," in Miriam K. Mills, ed., *Conflict Resolution and Public Policy* (New York: Greenwood, 1990), pp. 3–20; and Barry G. Rabe and John Martin Gillroy, "Intrinsic Value and Public Policy Choice: The Alberta Case," in John Martin Gillroy, ed., *Environmental Risk, Environmental Values, and Political Choices: Beyond Efficiency Tradeoffs in Public Policy Analysis* (Boulder, Colo.: Westview, 1993), pp. 150–70.

3. If anything, the very idea of hazardous waste and toxic substances may be most closely associated with past, well-publicized disasters stemming from their misuse and neglect. See Michael R. Reich, *Toxic Politics: Responding to Chemical Disasters* (Cornell University Press, 1991).

4. Andrew Blake, "Four Mass. Sites Picked by Waste Disposal Firm," *Boston Globe*, September 19, 1981, p. 13.

5. As Kent E. Portney noted, states make frequent revisions in their approaches to siting. "Consequently, it is somewhat difficult to provide an accurate and timely picture of where the states stand at any given time," he wrote. Kent E. Portney, *Siting Hazardous Waste Treatment Facilities: The Nimby Syndrome* (Westport, Conn.: Auburn House, 1991), p. 8. Among the more sophisticated efforts to chart these strategies at a given time, see Richard N. L. Andrews and Terrence K. Pierson, "Local Control or State Override: Experiences and Lessons to Date," *Policy Studies Journal*, vol. 14 (September 1985), pp. 90–99; and Richard N. L. Andrews and Philip Prete, "Trends in Hazardous Waste Facility Siting and Permitting," paper prepared for the March 1987 Conservation Foundation Workshop on Negotiating Hazardous Waste Facility Siting and Permitting Agreements.

6. David Osborne, *Laboratories of Democracy* (Harvard Business School Press, 1988).

7. William Gruber, "Solving the Equity Problem," *EI Digest: Industrial and Hazardous Waste Management* (July 1991), pp. 1–2.

8. Paul Langner, "Voluntary Approach to Hazardous Waste," *Boston Globe*, July 7, 1980, pp. 13, 27.

9. Bernd Holznagel, "Negotiation and Mediation: The Newest Approach to Hazardous Waste Facility Siting," *Boston College Environmental Affairs Law Review*, vol. 13, no. 3 (1986), pp. 354–55. For additional background on the Massachusetts approach to siting, see Lawrence S. Bacow and James R. Milkey, "Overcoming Local Opposition to Hazardous Waste Facilities: The Massachusetts Approach," *Harvard Environmental Law Review*, vol. 6, no. 2 (1982), pp. 265–305; and Michael O'Hare, Lawrence Bacow, and Debra Sanderson, *Facility Siting and Public Opposition* (Van Nostrand Reinhold, 1983).

10. Gail Bingham, *Resolving Environmental Disputes: A Decade of Experience* (Washington: Conservation Foundation, 1986); and Andrew S. McFarland, *Cooperative Pluralism: The National Coal Policy Experiment* (University of Kansas Press, 1993).

11. Geoffrey Castle, "Experiments in Hazardous Waste Facility Siting Legislation: The Massachusetts Experience," Working Paper (University of British Columbia, Sustainable Development Research Institute, February 1993), p. 24.

12. Gary Davis and Mary English, "Statutory and Legal Framework for Hazardous Waste Facility Siting and Permitting," paper prepared for the March 1987 Conservation Foundation Workshop on Negotiating Hazardous Waste Facility Siting and Permitting Agreements, p. 52.

13. Jerry Ackerman, "Waste Firm at Issue in Braintree Vote," *Boston Globe*, March 1, 1987, p. 32.

14. This and all subsequent unattributed quotes were from personal communications, in person and by telephone, with the author between May 1989 and June 1994. More than 150 individuals, representing government agencies, industry, and environmental groups, were interviewed on the condition that they would not be identified by name.

15. Andrew Blake, "Coalition Offers Aid for Hazardous Waste Facility," *Boston Globe*, October 23, 1981, pp. 17, 28; and Andrew Blake, "Sargent at Helm of Group Pushing for Waste Plants," *Boston Globe*, March 11, 1982, p. 27.

16. James P. Lester, for example, lists Massachusetts as one of fourteen states categorized as "progressives" on environmental regulation, indicating "a high commitment to environmental protection coupled with strong institutional capabilities." James P. Lester, "A New Federalism? Environmental Policy in the States," in Norman J. Vig and Michael E. Kraft, eds., *Environmental Policy in the 1990s*, 2d ed. (Washington: CQ Press, 1994), p. 63.

17. For a good overview of British Columbia politics and policy in the Social Credit era, see Michael Howlett and Keith Brownsey, "British Columbia: Public Sector Politics in a Rentier Resource Economy," in Keith Brownsey and Michael Howlett, eds., *The Provincial State: Politics in Canada's Provinces and Territories* (Mississauga, Ontario: Copp Clark Pitman, 1992), pp. 265–96.

18. British Columbia Waste Management Act of 1982 (S.B.C. 1982, c.41, as amended).

19. William M. Glenn, Deborah Orchard, and Thia M. Sterling, *Hazardous Waste Management Handbook*, 5th ed. (Don Mills, Ontario: Southam, 1988), British Columbia chapter, pp. 2–3.

20. Ibid., p. 3.

21. John Jackson and Phil Weller, *Chemical Nightmare* (Toronto, Ontario: Between the Lines, 1982), p. 110; and Gore and Storrie, Inc., *Canadian National Inventory of Hazardous Wastes* (Ottawa, Ontario: Environment Canada, 1982).

22. Glenn, Orchard and Sterling, *Hazardous Waste Management Handbook*, British Columbia chapter, p. 3.

23. Robert Paehlke and Douglas Torgerson, "Toxic Waste as Public Business," paper prepared for the 1990 annual meeting of the American Political Science Association, p. 7.

24. Ibid.

25. Richard N. L. Andrews, Raymond J. Burby, and Alvis G. Turner, *Hazardous Materials in North Carolina: A Guide for Decisionmakers in Local Government* (University of North Carolina Institute for Environmental Studies, 1985), pp. 120–24.

26. Frances M. Lynn, "Citizen Involvement in Hazardous Waste Sites: Two North Carolina Success Stories," *Environmental Impact Assessment Review*, vol. 7 (December 1987), p. 349.

27. Alice Tomboulian and Paul Tomboulian, "Perspectives on the Feasibility of Hazardous Waste Siting Using the Michigan Approach," *Great Lakes Waste and Pollution Review Magazine* (February 1984), pp. 1–6.

28. Robert D. Bullard, *Dumping in Dixie: Race, Class, and Environmental Quality* (Boulder, Colo.: Westview, 1990), chap. 3.

29. On the strategy of substituting one risk for another as a method to reduce opposition to facility siting, see Portney, *Siting Hazardous Waste Facilities*, chap. 7.

30. Letter from policy analyst Michael O'Hare to Mark Richards, quoted in Mark Richards, *Siting Industrial Facilities: Lessons from the Social Science Literature* (Washington: U.S. Council for Energy Awareness, 1992), p. 8.

31. Bruce A. Williams and Albert R. Matheny, *Democracy, Dialogue, and Social Regulation* (Yale University Press, forthcoming), chap. 2.

32. Ibid.

33. William T. Gormley, Jr., *Taming the Bureaucracy: Muscles, Prayers, and Other Strategies* (Princeton University Press, 1989).

34. Jean H. Peretz, "Hazardous Waste Defies Boundaries," *Environmental Management: PA Times 2d Annual Supplement* (July 1, 1991), p. 14.

35. Morell and Magorian, *Siting Hazardous Waste Facilities*.

36. Christopher Duerksen, *Environmental Regulation of Industrial Plant Siting: How to Make It Work Better* (Washington: Conservation Foundation, 1983).

37. Ontario Public Interest Research Group, *The South Cayuga Saga: A Chronological Account of the Happenings Associated with the Proposed Waste Disposal Site at South Cayuga* (Toronto, Ontario, 1981).

38. Press release issued by Donald Chant, Ontario Waste Management Corporation, November 18, 1981; Michael Scott, "South Cayuga II: The Role of the Ontario Waste Management Corporation," *Alternatives*, vol. 10 (Fall–Winter 1982), pp. 9–11; and Mario Ristoratore, "Siting Toxic Waste Disposal Facilities: A Comparative Study of Public Policy in the United States and Canada," Ph.D. dissertation, Brandeis University, 1986, chap. 4.

39. Ristoratore, "Siting Toxic Waste Disposal Facilities," p. 201.

40. Donald A. Chant, "Hazardous Waste Management in a Modern Society," paper prepared for the December 1985 Economic Council of Canada Colloquium on the Environment.

41. Ontario Waste Management Corporation, "Phase 4A Report Summary," (Toronto, Ontario, 1985). Much of the following information is derived from Robin Speis Norton, "Hazardous Waste Facility Siting in Canada: A Case Study of Ontario," University of Michigan School of Public Health, 1989. For a good overview of the initial three phases of the Ontario siting process, see Ontario Waste Management Corporation, *Facilities Development Process: Phase 3 Interim Report* (Toronto, Ontario, 1983). For a more general overview of hazardous waste management in Ontario, see Environmental Law Research Foundation, *Ontario Hazardous Waste Policy: A Provincial Forum* (Bolton, Ontario, 1986).

42. Glenn, Orchard, and Sterling, *Hazardous Waste Management Handbook*, Ontario chapter, pp. 7–8.

43. Norton, "Hazardous Waste Facility Siting in Canada."

44. Randy Turner, "Tough Sell: Waste Plant Scares Rural Towns," *Winnipeg Free Press*, February 24, 1990, p. 15.

45. Much of the subsequent discussion is influenced by Laura Flinchbaugh, "Hazardous Waste Facility Siting in Florida," University of Michigan School of Public Health, 1989. On the earlier development of hazardous waste policy in Florida, see Bruce A. Williams and Albert R. Matheny, "Hazardous Waste Policy in Florida: Is Regulation Possible?" in James P. Lester and Ann O'M. Bowman, eds., *The Politics of Hazardous Waste Management* (Duke University Press, 1983), pp. 74–101.

46. Florida Law, chap. 403.723 (1, 2).

47. "1992 Outlook for Commercial Hazardous Waste Management Facilities," *Hazardous Waste Consultant* (March–April 1992), p. 4.10.

48. Michael Heiman, "Using Public Authorities to Site Hazardous Waste Management Facilities: Problems and Prospects," *Policy Studies Journal*, vol. 18 (Summer 1990), p. 977.

49. Ristoratore, "Siting Toxic Waste Disposal Facilities," p. 186.

50. Heiman, "Using Public Authorities to Site Hazardous Waste Management Facilities," p. 978.

51. Barry G. Rabe and Janet B. Zimmerman, "Beyond Environmental Regulatory Fragmentation: Signs of Integration in the Case of the Great Lakes Basin," *Governance*, vol. 8 (January 1995).

52. North Carolina GS 143B–470.4(b). See also Andrews, Burby, and Turner, *Hazardous Materials in North Carolina*, pp. 125–26.

53. North Carolina GS 143B–470.4(b).

54. Tom Mather, "Frustrated Head of Waste Panel Calls It Quits," *Raleigh Times*, January 12, 1989. Much of the information on the Lee County case is derived from Ric Compton, "Hazardous Waste Siting in North Carolina," University of Michigan School of Public Health, 1991.

55. Memorandum from North Carolina state official George F. Givens to Committee on Finance, North Carolina General Assembly, "Senate Bill 324, Fourth Edition," May 18, 1989.

56. "EPA Threatens to Bar Waste Cleanup Funds for North Carolina," *Wall Street Journal*, March 15, 1991, p. B14.

57. Susan G. Hadden, *A Citizen's Right to Know: Risk Communication and Public Policy* (Boulder, Colo.: Westview, 1989).

58. Barry G. Rabe, "Environmental Regulation in New Jersey: Innovations and Limitations," *Publius: The Journal of Federalism*, vol. 21 (Winter 1991), pp. 83–103.

59. *New Jersey Hazardous Waste Facilities Plan*, prepared by Environmental Resources Management under contract to New Jersey Hazardous Waste Facility Siting Commission (Trenton, N. J., 1985); Mary English and Gary Davis, "American Siting Initiatives: Recent State Developments," in Bruce W. Piasecki and Gary A. Davis, eds., *America's Future in Toxic Waste Management: Lessons from Europe* (New York: Quorum, 1987), pp. 284–89; and Laura J. Pierce and Richard J. Gimello, "The Major Hazardous Waste Facilities Siting Act," *New Jersey Municipalities* (April 1989), pp. 8–9, 27–28.

60. Carl LaVo, "Not in My Backyard!" *National Wildlife*, vol. 26 (April–May 1988), p. 27.

61. Bob Narus, "Emotions Run High at Toxic-Wastes Site Hearings," *New York Times*, June 15, 1986, pp. 14–15.

Chapter 3

1. The province historically has emphasized subsidies to encourage the development of natural resources instead of any form of regulation related to environmental protection. Robert L. Mansell and Michael B. Percy, *Strength in Adversity: A Study of the Alberta Economy* (University of Alberta Press, 1990); and Larry Pratt, "The State and Province-Building: Alberta's Development Strategy," in Leo Panitch, ed., *The Canadian State: Political Economy and Political Power* (University of Toronto Press, 1977), pp. 133–62.

2. Peter J. Smith, "Alberta: A Province Just Like Any Other?" in Keith Brownsey and Michael Howlett, eds., *The Provincial State: Politics in Canada's Provinces and Territories* (Mississauga, Ontario: Copp Clark Pitman, 1992), pp. 243–64.

3. Michael Pretes, "Conflict and Cooperation: The Alaska Permanent Fund, the Alberta Heritage Fund, and Federalism," *American Review of Canadian Studies*, vol. 18 (Spring 1988), pp. 39–49.

4. Chem-Security Alberta, *Environmental Impact Assessment: Proposed Expansion of the Alberta Special Waste Treatment Centre* (June 1991), chap. 3.

5. For a survey of the numerous and varied estimates on waste generation in this period, see Geoffrey Castle, "Siting the Alberta Special Waste Treatment Centre: Minimum Self-Reproach, Maximum Community Choice," Discussion Paper (University of British Columbia, Sustainable Development Research Institute, 1992), pp. 2–3.

6. Alberta Hazardous Waste Management Committee, *A Summary of Inventory and Present Disposal Practices* (Edmonton, Alberta, 1979), pp. 1–2. This document drew heavily from a 270-page report by Reid Crowther and Partners that presented a disturbing picture of provincial waste disposal practices. It was so sweeping in its

condemnation of the prevailing order that it helped set the stage for the reforms that later developed.

7. Doug McConachie, "Time-Bomb Ticks in Dumps," *Edmonton Journal*, June 14, 1980, p. A1; and Doug McConachie, "Waste Monitoring Haphazard," *Edmonton Journal*, June 14, 1980, p. A3.

8. Glenn Bohn, "Where Waste Finds a Home . . . ," *Vancouver Sun*, November 22, 1986, p. B1.

9. Satya Das, "Gov't-Built Waste Plant Best — Citizens," *Edmonton Journal*, September 13, 1979, p. B2. Despite these two siting disasters, Kinetic Contaminants officials were outraged when the province shifted from a market strategy and developed the crown corporation and voluntary approach. See Allan Mayer, "Disposal Firm 'Shunted Aside,' " *Edmonton Journal*, January 24, 1981, p. B2.

10. Darcy Henton, "Hazardous Waste Plant Proposal Divides Town," *Edmonton Journal*, January 15, 1980, p. D10; and Don Thomas, "Hazardous Waste Site Studies for Two Hills," *Edmonton Journal*, December 27, 1979, p. B1.

11. On the use of direct democracy in the United States, see Thomas E. Cronin, *Direct Democracy: The Politics of Initiative, Referendum, and Recall* (Harvard University Press, 1989).

12. On the use of a direct democracy in Canada, see Patrick Boyer, *The People's Mandate: Referendums and a More Democratic Canada* (Toronto, Ontario: Dundurn Press, 1992).

13. Bruce A. Williams and Albert R. Matheny, *Democracy, Dialogue, and Social Regulation* (Yale University Press, forthcoming).

14. For a more detailed overview of this process, see Jennifer McQuaid-Cook and Kenneth J. Simpson, "Siting a Fully Integrated Waste Management Facility in Alberta," *Journal of the Air Pollution Control Association*, vol. 36 (September 1986), pp. 1031–36; and R. C. MacKenzie, "Canadian City Welcomes Hazwaste Facility," *World Wastes* (July 1984), pp. 46–50.

15. Barb Livingstone, "Hazardous-Waste Plants Needed," *Edmonton Journal*, September 28, 1981, p. A6.

16. Karen Booth, "Beaver County Petition Backs Waste Plant Site," *Edmonton Journal*, April 4, 1982, p. A13; Ed Struzik, "Scare Tactics Charged in Waste-Plant Plebiscite," *Edmonton Journal*, April 13, 1982, p. B1; and Ed Struzik, "Beaver County Voters Say No to Waste Plant," *Edmonton Journal*, April 15, 1982, p. B8.

17. Castle, "Siting the Alberta Special Waste Treatment Centre," p. 15.

18. William T. Gormley, Jr., *Taming the Bureaucracy: Muscles, Prayers, and Other Strategies* (Princeton University Press, 1989), p. 77. See also Daniel A. Mazmanian and Jeanne Nienaber, *Can Organizations Change? Environmental Protection, Citizen Participation, and the Corps of Engineers* (Brookings, 1979).

19. Kent E. Portney provides further warning against the potential limitations of public participation measures in fostering siting agreements: "There is plenty of evidence to suggest that participation processes might be effective, indeed might be needed, in order to build the kind of long-term trust and personal capacity to make facility siting possible. Yet, as envisioned to date, inclusionary approaches which focus only on short-term and sporadic participation are likely to meet with every bit as much difficulty as their exclusionary counterparts." Kent E. Portney, *Siting Hazardous Waste*

Treatment Facilities: The Nimby Syndrome (Westport, Conn.: Auburn House, 1991), p. 68.

20. See Gormley, *Taming the Bureaucracy*; Robert O. Keohane, *After Hegemony: Cooperation and Discord in the World Political Economy* (Princeton University Press, 1984); Alan Stone, *Public Service Liberalism: Telecommunications and Transitions in Public Policy* (Princeton University Press, 1991); David Osborne and Ted Gaebler, *Reinventing Government: How the Entrepreneurial Spirit Is Transforming the Public Sector* (Reading, Mass.: Addison-Wesley, 1992); and Oran R. Young, *International Cooperation: Building Regimes for Natural Resources and the Environment* (Cornell University Press, 1989).

21. On the history of Canadian crown corporations and the ways in which they operate, see Douglas F. Stevens, *Corporate Autonomy and Institutional Control: The Crown Corporation as a Problem in Organization Design* (McGill-Queen's University Press, 1993); and Jeanne Kirk Lauz and Maureen Appel Molot, *State Capitalism: Public Enterprise in Canada* (Cornell University Press, 1988). The Stevens book is particularly strong on discussing the evolution of crown corporations in Alberta and Manitoba.

22. Alberta Hazardous Waste Team, *Hazardous Wastes in Alberta* (Edmonton: Alberta Environment, 1981).

23. R. Kent Weaver and Bert A. Rockman, eds., *Do Institutions Matter? Government Capabilities in the United States and Abroad* (Brookings, 1993).

24. Barry G. Rabe, *Fragmentation and Integration in State Environmental Management* (Washington: Conservation Foundation, 1986).

25. Barry G. Rabe, "Cross-Media Environmental Regulatory Integration: The Case of Canada," *American Review of Canadian Studies*, vol. 19 (Autumn 1989), pp. 261–74.

26. On the definition of redistributive policy and an analysis of the complexity of program implementation in this area, see Paul E. Peterson, Barry G. Rabe, and Kenneth K. Wong, *When Federalism Works* (Brookings, 1986).

27. For a more general discussion of the politics that stem from such a combination of cost and benefit concentration, see James Q. Wilson, *Bureaucracy: What Government Agencies Do and Why They Do It* (Basic Books, 1989), chap. 5.

28. In many programs that provide professional education for future environmental and hazardous waste policy professionals, the teaching of technical skills of risk assessment and risk management is not integrated with those of risk communication. Consequently, hazardous waste policy professionals possess extensive technical training but are largely ignorant of the social and political setting in which those skills must be used. This is evident in both the private and public sectors.

29. Natalia M. Krawetz, *Hazardous Waste Management: A Review of Social Concerns and Aspects of Public Involvement* (Edmonton: Alberta Environment Research Secretariat, 1979), p. 10.

30. William M. Glenn, Deborah Orchard, and Thia M. Sterling, *Hazardous Waste Management Handbook*, 5th ed. (Don Mills, Ontario: Southam, 1988), Alberta chapter, pp. 4–5; "Concern over Turnover," *Swan Hills Grizzly Gazette*, January 15, 1986, p. 1; and "Conscience Demands Resignation," *Swan Hills Grizzly Gazette*, February 5, 1986, p. 6.

31. Ian Williams, "Swan Hills Still Carries Transient Tag," *Edmonton Journal*, January 4, 1978; and Kelly McParland, "Oil Cuts Drain Town's Lifeblood," *Edmonton Journal*, July 10, 1981, p. B1.

32. Len Stahl, "$50 Million Plant Generates Opportunity," *Trade and Commerce Magazine*, vol. 84 (December 1989), pp. 224–25.

33. Lasha Morningstar, "Dumping Our Hazardous Waste Outside Alberta 'Irresponsible,'" *Edmonton Journal*, September 22, 1979, p. A1.

34. "Hazardous Waste a Question of Dump Sharing — Cookson," *Edmonton Journal*, April 23, 1981, p. B6; and "Alta. No Dumping Ground for Wastes, Says Cookson," *Edmonton Journal*, April 22, 1981, p. B2

35. Natalia M. Krawetz, "Social Concerns," Edmonton, Alberta, Hazardous Waste Management Committee, 1980, pp. 1–2.

36. Doug Macdonald, *The Politics of Pollution: Why Canadians Are Failing Their Environment* (Toronto, Ontario: McClelland and Stewart, 1991), p. 221; and Glenn, Orchard, and Sterling, *Hazardous Waste Management Handbook*, Alberta chapter, p. 7.

37. Miro Cernetig, "Alberta May Lift Barrier to Waste," *Toronto Globe and Mail*, June 8, 1990; "NRCB Hearings on Expanding the Market Area for the Alberta Special Waste Treatment Centre," *In Our Backyard*, vol. 4 (Fall 1993), p. 2; and "NRCB Hearings on Proposed Policy Changes," *In Our Backyard*, vol. 5 (Winter 1994), pp. 1–2.

38. Avis Schleske, "ASWTC Expansion Public Meeting Held," *Swan Hills Grizzly Gazette*, July 9, 1991, p. 1.

39. Chem-Security Alberta, *Environmental Impact Assessment*.

40. Glenn, Orchard, and Sterling, *Hazardous Waste Management Handbook*, Alberta chapter, pp. 8–9.

41. "Ryley Breaks New Ground," *In Our Backyard*, vol. 1 (October 1990), p. 1.

42. Chris Zeiss, "Directions for Engineering Contributions to Successfully Siting Hazardous Waste Facilities," paper prepared for the 1993 conference on Siting Hazardous Waste Facilities: Policies and Approaches for the 1990s, University of British Columbia; and Chris Zeiss, "Community Decisionmaking and Impact Management Priorities for Siting Waste Facilities," *Environmental Impact Assessment Review*, vol. 11 (September 1991), pp. 231–57.

43. Glenn, Orchard, and Sterling, *Hazardous Waste Management Handbook*, Alberta chapter, pp. 8–9.

44. "For Syncrude at Fort McMurray, Waste Minimization Is a Refined Process," *In Our Backyard*, vol. 2 (December 1991), pp. 2–3.

45. "Alberta's Great Drug Round Up," *In Our Backyard*, vol. 4 (Spring 1993), pp. 1–2.

46. "Speaking of Round Ups . . . ," *In Our Backyard*, vol. 2 (Summer 1991), p. 3.

47. For an overview of this research, see Peter K. Eisinger, *The Rise of the Entrepreneurial State: State and Local Economic Development Policy in the United States* (University of Wisconsin Press, 1989), pp. 169–72. See also Christopher Duerksen, *Environmental Regulation of Industrial Plant Siting: How to Make It Work Better* (Washington: Conservation Foundation, 1983); and Stephen M. Meyer, "Environmentalism and Economic Prosperity: Testing the Environmental Impact Hypothesis," Massachusetts Institute of Technology, Project on Environmental Politics and Policy, October 1992.

48. Barry G. Rabe, "Environmental Regulation in New Jersey: Innovations and Limitations," *Publius: The Journal of Federalism*, vol. 21 (Winter 1991), pp. 83–103.

Chapter 4

1. The classic contributions to this vast literature include Jack L. Walker, Jr., "The Diffusion of Innovations among the American States," *American Political Science Review*, vol. 63 (September 1969), pp. 880–99; and Virginia Gray, "Innovation in the States: A Diffusion Study," *American Political Science Review*, vol. 67 (December 1973), pp. 1174–85.

2. Manitoba Hazardous Waste Management Corporation, *Hazardous Waste Market Characterization Study for Manitoba* (Winnipeg, Manitoba, 1989), p. i. On earlier disposal practices, see also Edwin Yee, Emil Kucera, and J. J. Keleher, eds., *Hazardous Waste Management in Manitoba* (Winnipeg, Manitoba: Manitoba Environment and Workplace Safety and Health, 1985).

3. Manitoba Hazardous Waste Management Corporation, *Fourth Annual Report 1990* (Winnipeg, Manitoba, 1990).

4. Yee, Kucera, and Keleher, eds., *Hazardous Waste Management in Manitoba*, p. 13.

5. Estimates of the total number of public meetings vary somewhat but consistently indicate a level of public outreach comparable only to Alberta among other Canadian provinces and American states. See Randy Turner, "Tough Sell: Waste Plant Scares Rural Towns," *Winnipeg Free Press*, February 24, 1990, pp. 15–22; and Manitoba Hazardous Waste Management Corporation, *Third Annual Report 1989* (Winnipeg, Manitoba, 1989), pp. 12–13.

6. John Lyons, "Waste Plants Draws Towns," *Winnipeg Free Press*, May 23, 1989, p. 2; and Manitoba Hazardous Waste Management Corporation, *Third Annual Report 1989*, p. 5.

7. Randy Turner, " 'Bunch of Ukrainians' Say No to Hazardous Waste Site," *Winnipeg Free Press*, January 18, 1990, pp. 1, 4; Randy Turner, "Townsfolk Wary of Divisive Issue," *Winnipeg Free Press*, January 23, 1990, p. 22; "The Place for Waste Disposal," *Winnipeg Free Press*, January 20, 1990, p. 18; and Manitoba Hazardous Waste Management Corporation, *Fourth Annual Report 1990*, p. 9.

8. Maureen Houston, "Pinawa Fearless about Disposal Site," *Winnipeg Free Press*, November 2, 1989, p. 2.

9. Janet McFarland, "City Suggested as Disposal Site," *Winnipeg Free Press*, January 19, 1990, p. 1; "The Place for Waste Disposal"; and Manitoba Hazardous Waste Management Corporation, *Fourth Annual Report 1990*, p. 10.

10. Geoffrey Castle, "Siting Hazardous Waste Treatment Facilities in Western Canadian Provinces: 1980–1993," paper prepared for the 1993 conference on Siting Hazardous Waste Facilities: Policies and Approaches for the 1990s, University of British Columbia, pp. 21–22.

11. Ibid.

12. Manitoba Hazardous Waste Management Corporation, *Second Annual Report 1988* (Winnipeg, Manitoba, 1988), p. 11.

13. "Hazardous Waste Study Continues," *Winnipeg Free Press*, April 8, 1990, p. 8.

14. Dennis McKnight 2051 Inc., *Quantitative Investigation of the Attitudes and Perceptions of the Residents of the R. M. of Montcalm towards a Hazardous Waste Treatment Facility* (Winnipeg, Manitoba: Manitoba Hazardous Waste Management Corporation, 1991).

15. For an excellent overview of the latter stages of this process, see Geoffrey Castle, "Hazardous Waste Facility Siting in Manitoba: Case Study of a Success," Sustainable Development Research Institute, Discussion Paper (University of British Columbia, 1992).

16. Rural Municipality of Montcalm and Manitoba Hazardous Waste Management Corporation Co-Management Agreement (1992).

17. "Co-Management," *Waste Paper*, vol. 4 (Winnipeg, Manitoba: Manitoba Hazardous Waste Management Corporation, 1992), p. 3.

18. Manitoba Hazardous Waste Management Corporation, *Sixth Annual Report 1992* (Winnipeg, Manitoba, 1992), p. 3.

19. This may reflect a reluctance to tamper with a seemingly successful project that offered long-term economic development advantages if implemented. It may further reflect a more incremental period of change during the years of the late 1980s and early 1990s, when various parties shifted in and out of power. As Alex Netherton noted, even after winning a majority in 1990, "The Conservatives had pledged to hold on to the moderate neo-liberal strategy associated with the minority government period. There has been no attempted exorcisms of significant policy innovations. This would suggest that there is neither the political support for nor the capacity to undertake paradigmatic change." Alex Netherton, "Manitoba: The Shifting Points of Politics, A Neo-Institutional Analysis," in Keith Brownsey and Michael Howlett, eds., *The Provincial State: Politics in Canada's Provinces and Territories* (Mississauga, Ontario: Copp Clark Pitman, 1992), p. 200.

20. According to Paul G. Thomas, "While Manitoba's diversified economy weathers the highs and lows of fluctuating economic cycles well, it lacks the dynamic sectors that stimulate strong and sustained growth." Paul G. Thomas, "Manitoba: Stuck in the Middle," in Ronald L. Watts and Douglas M. Brown, eds., *Canada: The State of the Federation 1989* (Kingston, Ontario: Institute of Intergovernmental Relations, 1989), p. 78.

21. Such analysts call for governmental strategies that are entrepreneurial, citizen-responsive, mission-driven, and results-oriented. The best known work of this genre is David Osborne and Ted Gaebler, *Reinventing Government: How the Entrepreneurial Spirit Is Transforming the Public Sector* (Reading, Mass.: Addison-Wesley, 1992). For a sophisticated interpretation of this literature, see John J. DiIulio, Jr., Gerald Garvey, and Donald F. Kettl, *Improving Government Performance: An Owner's Manual* (Brookings, 1993).

22. Castle, "Hazardous Waste Facility Siting in Manitoba," p. 23.

23. Donald Campbell, "Corporation Reveals Details Except Location," *Winnipeg Free Press*, April 1, 1990, p. 1.

24. Randy Turner, "Hamiota Sets June for Vote on Waste Centre: Feuding Factions Pan Ballot Date," *Winnipeg Free Press*, March 30, 1990, p. 2.

25. "Economic Effects," *Waste Paper*, vol. 4 (Winnipeg, Manitoba: Manitoba Hazardous Waste Management Corporation, 1992), p. 6.

26. Rural Municipality of Montcalm and Manitoba Hazardous Waste Management Corporation Co-Management Agreement.

27. Ibid.

28. Manitoba Hazardous Waste Management Corporation, *Sixth Annual Report 1992*, p. 4.

29. Ibid.

30. "Toxic Waste in Our Back Yard?" *Selkirk Enterprise and Lake Centre News*, October 27, 1986, p. 4.

31. "Ryz Replies," *Selkirk Enterprise and Lake Centre News*, November 3, 1986, p. 5.

32. Rural Municipality of Montcalm and Manitoba Hazardous Waste Management Corporation Co-Management Agreement.

33. Graham P. Latonas, *Proposed Site Selection Process for the Development of a Hazardous Waste Management System* (Winnipeg, Manitoba: Manitoba Hazardous Waste Management Corporation, 1988), pp. 23–24.

34. Manitoba Hazardous Waste Management Corporation, *Fourth Annual Report 1990*, p. 8.

35. Dave Witty, quoted in "Healthy Communities Projects," *Waste Paper*, vol. 3 (Spring 1991), p. 2.

36. Michael Heiman, "Using Public Authorities to Site Hazardous Waste Management Facilities: Problems and Prospects," *Policy Studies Journal*, vol. 18 (Summer 1990), p. 981.

37. Barbara Connell and Ross Edmunds, "A Waste Management Success Story: Gravure Graphics," *Waste Paper*, vol. 3 (Spring 1991), p. 3.

38. William Gruber, "North Carolina's Hazardous Waste: Pursuing a 'We Make It, You Take It' Policy," *EI Digest: Industrial and Hazardous Waste Management* (September 1991), p. 16.

39. Jonathan Walters, "The Poisonous War over Hazardous Waste," *Governing*, vol. 5 (November 1991), p. 32.

40. John Hodges-Copple, "State Roles in Siting Hazardous Waste Facilities," *Forum for Applied Research and Public Policy*, vol. 2 (Fall 1987), p. 85; and Monte Basgell, "A Tale of Two Cities and Hazardous Waste," *Raleigh News and Observer*, October 20, 1985.

41. "1989 Outlook for Commercial Hazardous Waste Management Facilities A Nationwide Perspective," *Hazardous Waste Consultant* (March–April 1989), pp. 4–33.

42. Portions of the following discussion are drawn from Frances M. Lynn's excellent case study of the ECOFLO proposal. See Frances M. Lynn, "Citizen Involvement in Hazardous Waste Sites: Two North Carolina Success Stories," *Environmental Impact Assessment Review*, vol. 7 (December 1987), p. 356.

43. Quoted in ibid., p. 355.

44. Interview with Tom Barbee on Greensboro television Channel 8, July 14, 1984, quoted in ibid.

45. For additional information on this grant and the three North Carolina task forces in Guilford, Durham, and Mecklenburg that it supported, see Richard N. L. Andrews, Raymond J. Burby, and Alvis G. Turner, *Hazardous Materials in North Caro-*

lina: A Guide for Decisionmakers in Local Government (University of North Carolina Institute for Environmental Studies, 1985), pp. 25–27.

46. Ibid., p. 26.

47. Ibid., pp. 27–28.

48. Ibid., pp. 28–29.

49. Lynn, "Citizen Involvement in Hazardous Waste Sites," pp. 355–56.

50. Hodges-Copple, "State Roles in Siting Hazardous Waste Facilities," pp. 85–86.

51. Quoted in Lynn, "Citizen Involvement in Hazardous Waste Sites," p. 349.

52. In 1987, for example, of the 38,100 tons of hazardous waste generated in Minnesota that could not be managed on-site, 29,800 tons were sent out of state, with the remainder placed in Minnesota landfills. Minnesota Waste Management Board, *Stabilization and Containment Report on Facility Development* (St. Paul, Minn., 1988), p. 35.

53. Dan Reinke, "Development of a Stabilization and Containment Facility in Minnesota," paper prepared for the 1988 annual meeting of the Air Pollution Control Association, pp. 1–2; and Richard N. L. Andrews and Terrence K. Pierson, "Hazardous Waste Facility Siting Processes: Experience from Several States," *Hazardous Waste*, vol. 1, no. 3 (1984), p. 380.

54. Bobbie Wilkering, "Koochiching Turns Down Hazardous Waste Faculty," *Bemidji Pioneer*, March 22, 1989, p. 1A; Laurel Beager, "County Withdraws from Waste Siting," *International Falls Daily Journal*, March 22, 1989, p. 1; and Marsha Shoemaker, "Koochiching Drops Waste Site Bid," *Grand Forks Herald*, March 22, 1989, p. 1A.

55. Potential financial rewards became far greater for counties that agreed to remain active well into the process. For example, any county that remained involved at the stage at which environmental impact statements would be completed, indicating some form of local support for an agreement, would receive $150,000 a year for up to two years. These funds could be used "for any area of county government." See Reinke, "Development of a Stabilization and Containment Facility in Minnesota," pp. 3–4.

56. Todd Anderson, "Possible Host for Waste Site Down to Three Counties," *Red Lake Falls Gazette*, August 24, 1988, p. 1.

57. Charles Laszewski, "Waste Board Chief Says Mistrust Led to Firing," *St. Paul Pioneer Press Dispatch*, August 30, 1988, p. 1B; and Charles Laszewski, "Review of Waste Board Firings Demanded," *St. Paul Pioneer Press Dispatch*, September 8, 1988, p. 1B.

58. Bruce Orwall, "Audit May Stall Bill Reviving Waste Board," *St. Paul Pioneer Press Dispatch*, April 28, 1989, p. 1B; and Laurel Beager, "Audit Says WMB Misused Funds in Siting Process," *International Falls Daily Journal*, May 11, 1989, p. 1.

59. Reinke, "Development of a Stabilization and Containment Facility in Minnesota," p. 10.

60. For a full discussion of the option of a facility owned by the state and managed by a private firm on contract, see Minnesota Waste Management Board, *Stabilization and Containment Report on Facility Development* (St. Paul, Minn., 1988), pp. 11–16.

61. Ibid., p. 13.

62. Contract between the State of Minnesota and Red Lake County for the Location, Development, Operation, Closure and Care of a Facility for the Stabilization and

Containment of Hazardous Wastes (September 20, 1990), article 20, pp. 70–80, and article 16, p. 52.

63. See, for example, Minnesota Office of Waste Management, *Report on Barriers to Pollution Prevention* (St. Paul, Minn., 1991).

64. For a more thorough review of these programs, see U.S. Council on Environmental Quality, *Environmental Quality: The Twentieth Annual Report of the Council on Environmental Quality* (Washington, 1990).

65. Cindy A. McComas, "Minnesota Technical Assistance Program: Waste Reduction Assistance for Small Quantity Hazardous Waste Generators," *Hamline Journal of Public Law and Policy*, vol. 8 (1986), pp. 497–506; and Roberta G. Gordon, "Legal Incentives for Reduction, Reuse, and Recycling: A New Approach to Hazardous Waste Management," *Yale Law Journal*, vol. 95 (1986), pp. 810–31.

66. On the problems of medium-based regulatory fragmentation at the state level and more integrative options, see Barry G. Rabe, *Fragmentation and Integration in State Environmental Management* (Washington: Conservation Foundation, 1986).

67. The Minnesota Toxic Pollution Prevention Act of 1990, secs. 7 and 8, chap. 115D.

68. Barry G. Rabe, *Integrated Permitting: Experience and Innovation at the State Level*, report submitted to the Environment Program of the Office of Technology Assessment, U.S. Congress (July 1994).

69. John Chell, "3Rs of Sustainable Development," *Resource* (July–August 1993), p. 6.

70. On the earlier history of California hazardous waste policy, see Daniel A. Mazmanian, Michael Stanley-Jones, and Miriam J. Green, *Breaking Political Gridlock: California's Experiment in Public-Private Cooperation for Hazardous Waste Policy* (Claremont, Calif.: California Institute of Public Affairs, 1988), chaps. 1–3; David L. Morell, "Technological Policies and Hazardous Waste Politics in California," in James P. Lester and Ann O'M. Bowman, eds., *The Politics of Hazardous Waste Management* (Duke University Press, 1983), pp. 139–75; and David L. Morell, "The Siting of Hazardous Waste Facilities in California," *Public Affairs Report*, vol. 25 (October 1984), pp. 1–10.

71. "County Subject to More Than Its 'Fair Share' of Hazardous Waste, Sierra Club Suit Charges," *Environmental Reporter*, July 21, 1989, pp. 582–83.

72. David L. Morell, "Hazardous Waste Management and Facility Siting in California," paper prepared for the 1993 conference on Siting Hazardous Waste Facilities: Policies and Approaches for the 1990s, University of British Columbia, p. 35.

73. Morell, "Hazardous Waste Management and Facility Siting in California."

74. Ibid., p. 5.

75. Ibid., p. 13; and "County Subject to More Than Its 'Fair Share' of Hazardous Waste," pp. 582–83.

76. Morell, "Hazardous Waste Management and Facility Siting in California," p. 15.

77. Speaking for many officials in states that "import" large quantities of California hazardous waste, Utah House Speaker H. Craig Moody contended, "California is making us their dumping ground." See Charles McCoy, "California Toxic Waste Exports Draw Ire," *Wall Street Journal*, July 2, 1991, p. B1.

78. Environmental Information, *Interdependence of the Management of Hazardous Waste* (St. Paul, Minn., 1993), sec. 5.

79. Daniel A. Mazmanian and David L. Morell, "The Nimby Syndrome: Facility Siting," in Norman J. Vig and Michael E. Kraft, eds., *Environmental Policy in the 1990s*, 2d ed. (Washington: CQ Press, 1994), pp. 245–46; and Daniel A. Mazmanian and David L. Morell, *Beyond Superfailure: America's Toxics Policy for the 1990s* (Boulder, Colo.: Westview, 1992), pp. 202–03.

80. For a more detailed review of the Quebec case, see Barry G. Rabe and Pamela I. Protzel, "The Pitfalls of Decentralized Environmental Regulation: The Case of Hazardous Waste Policy in Quebec," University of Michigan School of Public Health, 1991.

81. One of the most thorough analyses of Quebec's problems was published by the Group for Study and Restoration of Hazardous-Waste Disposal Sites in 1984. It concluded that at least 300 major abandoned hazardous waste dumps were in the province and that at least 12,000 industries were contributing to their contamination. The report also highlighted three additional problems: the unauthorized storage of waste; the disposal of waste in unauthorized sites; and the absence of accurate records on the amount of waste produced or deposited. For a useful summary, see Marcel Gaucher, "Managing Sites Contaminated by Hazardous Substances in Quebec," in *Reclamation, Conservation, and the Environment* (Ottawa, Ontario: Energy, Mines, and Resources Canada, 1988), pp. 2–7.

82. William M. Glenn, Deborah Orchard, and Thia M. Sterling, *Hazardous Waste Management Handbook*, 5th ed. (Don Mills, Ontario: Southam, 1988), Quebec chapter, p. 5.

83. Mario Ristoratore, "Siting Toxic Waste Disposal Facilities: A Comparative Study of Public Policy in the United States and Canada," Ph.D. dissertation, Brandeis University, 1986, p. 76.

84. Peter Ward, "The Town That Took Poison," *International Wildlife*, vol. 15 (March–April 1985), p. 33.

85. Ristoratore, "Siting Toxic Disposal Facilities," pp. 71–72.

86. Christopher Scanlan, "Hazardous Waste Crosses the Border," *Detroit Free Press*, May 15, 1991, p. 8A.

87. Rick Boychuk, "Many Quebec Firms Not Using Waste-Treatment Plant," *Montreal Gazette*, December 1, 1987, p. A1.

88. Commission D'Enquête sur les Déchets Dangereux, *Hazardous Waste in Quebec: Issues and Questions* (Montreal, Quebec, 1989), p. 11.

89. Barry G. Rabe, "Exporting Hazardous Waste in North America," *International Environmental Affairs*, vol. 3 (Spring 1991), pp. 108–23.

Chapter 5

1. Low-level wastes include materials contaminated with small amounts of radioactive substances, such as clothing, packaging, animal carcasses, medical fluids, and power reactor liquids. They need to be isolated from humans for between 60 and 300 years. High-level wastes include the spent fuel from nuclear reactors and the concentrated fission products left in liquid form after spent fuel has been chemically reproc-

essed to retrieve plutonium and unfissioned uranium. See Barry G. Rabe, "Low-Level Radioactive Waste Disposal and the Revival of Environmental Regionalism in the United States," *Environment and Planning Law Journal*, vol. 7 (September 1990), pp. 171–80.

2. For an excellent overview of the American case, see Mary R. English, *Siting Low-Level Radioactive Waste Disposal Facilities: The Public Policy Dilemma* (New York: Quorum, 1992); and Riley E. Dunlap, Michael E. Kraft, and Eugene A. Rosa, eds., *Public Reactions to Nuclear Waste: Citizens' Views of Repository Siting* (Duke University Press, 1993).

3. Doug Macdonald, *The Politics of Pollution: Why Canadians Are Failing Their Environment* (Toronto, Ontario: McClelland and Stewart, 1991), chap. 14.

4. John Urquhart, "Canada Nuclear Industry Expects Orders to Rise, Plans Compact Reactor for U.S.," *Wall Street Journal*, February 10, 1990, p. B6.

5. Robert P. Stoker, *Reluctant Partners: Implementing Federal Policy* (University of Pittsburgh Press, 1991), pp. 158–59.

6. English, *Siting Low-Level Radioactive Waste Disposal Facilities*.

7. Canadian Mining and Engineering Corporation, *Port Granby: Waste Management Facility* (Toronto, 1987), p. 1.

8. Canadian Ministry of Energy, Mines, and Resources, *Opting for Cooperation: Report of the Siting Process Task Force on Low-Level Radioactive Waste Disposal* (Ottawa, Ontario: Energy, Mines, and Resources Canada, 1987), p. 3.

9. Ibid., p. ix.

10. Canadian Ministry of Energy, Mines, and Resources, *Opting for Cooperation: The First Phases* (Ottawa, Ontario: Energy, Mines, and Resources Canada, 1990), p. 3.

11. Ibid., pp. 13, 14.

12. Macdonald, *Politics of Pollution*, chap. 14.

13. On the remedial action process of the International Joint Commission, see John H. Hartig and Michael A. Zarull, eds., *Under RAPs: Toward Grassroots Ecological Democracy in the Great Lakes Basin* (University of Michigan Press, 1992).

14. Ibid., and Barry G. Rabe and Janet Zimmerman, "Cross-Media Environmental Integration in the Great Lakes Basin," *Environmental Law*, vol. 22, no. 1 (1992), pp. 253–79.

15. Ray Kemp, *The Politics of Radioactive Waste Disposal* (Manchester University Press, 1992).

16. The idea of such regional approaches to pressing problems, including environmental ones, has been widely used in the United States. See Martha Derthick, *Between State and Nation: Regional Organizations of the United States* (Brookings, 1974). On regional approaches to environmental management in Canada, see M. Paul Brown, "Environment Canada and the Pursuit of Administrative Decentralization," *Canadian Public Administration*, vol. 29 (Summer 1986), pp. 218–36.

17. On the early history of this legislation, see Richard C. Kearney and John J. Stucker, "Interstate Compacts and the Management of Low-Level Radioactive Wastes," *Public Administration Review*, vol. 45 (January–February 1985), pp. 210–20; E. William Colglazier, Jr., ed., *The Politics of Nuclear Waste* (Pergamon, 1982); and Richard C. Kearney and Robert B. Garey, "American Federalism and the Management of Radioactive Wastes," *Public Administration Review*, vol. 42 (January–February 1982), pp. 14–24.

18. *New York v. United States,* 1125 S. Ct. 2408 (1992); and Paul M. Barnett, "New York State Is Key Victor in Waste Suit," *Wall Street Journal,* June 22, 1992, p. A4.

19. Fred Thomas, "Claims Widen Rift over Initiative 402," *Omaha World-Herald,* November 9, 1988, p. A3.

20. A leader of this organization went on a thirty-day hunger strike to draw media attention in early 1990. Kent Warneke, "Dump Foes Hear Options," *Norfolk Daily News,* March 21, 1990, p. 1.

21. This shift in policy was approved by the Connecticut legislature in April 1993. It proposed extensive public participation provisions combined with financial support to potential volunteers. See Richard C. Kearney and Ande A. Smith, "The Low-Level Radioactive Waste Siting Process in Connecticut: Anatomy of a Failure," paper prepared for the 1993 annual meeting of the American Society for Public Administration, pp. 25–26.

Chapter 6

1. See, for example, Bruce W. Piasecki and Wendy Grieder, *America's Future in Toxic Waste Management: Lessons from Europe* (New York: Quorum, 1987); William Mangun, "A Comparative Analysis of Hazardous Waste Management Policy in Western Europe," in Charles E. Davis and James P. Lester, eds., *Dimensions of Hazardous Waste Politics and Policy* (New York: Greenwood, 1988), pp. 205–21; and Joanne Linnerooth and Allen V. Kneese, "Hazardous Waste Management: A West German Approach," *Resources,* no. 96 (Summer 1989), pp. 7–10. On the European experience in low-level radioactive waste facility siting and management, see Ray Kemp, *The Politics of Radioactive Waste Disposal* (Manchester University Press, 1992).

2. For a more precise definition of cooperative and other varieties of federalism, see David W. Welborn, "Conjoint Federalism and Environmental Regulation in the United States," *Publius: The Journal of Federalism,* vol. 18 (Winter 1988), pp. 28–43. Although this analysis is applied to the United States, similar terminology is used in the analysis of intergovernmental relations in Canada.

3. Ibid. For further analysis of the American approach to regulatory federalism, see U.S. Advisory Commission on Intergovernmental Relations, *Regulatory Federalism: Policy, Process, Impact, and Reform* (Washington, 1984); and Thomas J. Anton, *American Federalism and Public Policy: How the System Works* (Temple University Press, 1989), chap. 8.

4. James A. Morone, *The Democratic Wish: Popular Participation and the Limits of American Government* (Basic Books, 1990).

5. Thomas E. Cronin, *Direct Democracy: The Politics of Initiative, Referendum, and Recall* (Harvard University Press, 1989).

6. Patrick Boyer, *The People's Mandate: Referendums and a More Democratic Canada* (Toronto, Ontario: Dundurn Press, 1992); and A. Paul Pross, *Group Politics and Public Policy* (Oxford University Press, 1986).

7. Bunyan Bryant and Paul Mohal, eds., *Race and the Incidence of Environmental Hazards* (Boulder, Colo.: Westview, 1992).

8. In the most prominent of these cases, the U.S. Supreme Court in June 1992 reversed an earlier Alabama Supreme Court decision to uphold Alabama import re-

strictions, which had been devised primarily to reduce the enormous flow on out-of-state waste headed to the massive Emelle facility each year. In its ruling, the U.S. Supreme Court concluded that the "additional fee" that Alabama imposed on imported waste violated the commerce clause of the U.S. Constitution. It emphasized that "some reason, apart from origin" must exist for treating imported wastes differently, noting that any public health hazards imposed by these wastes were no different whether they were generated in Alabama or elsewhere. In his lone dissent, Chief Justice William H. Rehnquist emphasized historic state powers to control their borders when facing a public health threat, as is typical in a quarantine situation. This decision and other complementary ones in 1992 confirmed a line of judicial interpretation initiated in 1978. While they did not resolve all aspects of the issue, they clearly rejected a series of restrictive strategies that Alabama and more than two dozen other states had developed in the 1980s. *Chemical Waste Management, Inc.* v. *Guy Hunt*, 121 S. Ct. 2009 (1992). See also *New York* v. *United States*, 1125 S. Ct. 2408 (1992). For a concise summary of these cases, see "Intergovernmental Digest," *Intergovernmental Perspectives*, vol. 18 (Summer 1992), pp. 14–16.

9. Quoted in William Gruber, "Solving the Equity Problem," *EI Digest: Industrial and Hazardous Waste Management* (July 1991), p. 3.

10. Ibid.; and Jonathan Walters, "The Poisonous War over Hazardous Waste," *Governing*, vol. 5 (November 1991), pp. 32–35.

11. On the role of these organizations and additional ones in the integration of environmental regulatory programs, see Frances H. Irwin, "An Integrated Framework for Preventing Pollution and Protecting the Environment," *Environmental Law*, vol. 22, no. 1 (1992), pp. 52–62. See also Pollution Probe, *Profit from Pollution Prevention: A Guide to Industrial Waste Reduction and Recycling*, 2d ed. (Toronto, Ontario, 1991).

12. Frank Edward Allen, "Reducing Toxic Waste Produces Quick Results," *Wall Street Journal*, August 11, 1992, p. B1.

13. For a more detailed discussion of the strengths and weaknesses of these approaches, see Daniel A. Mazmanian and David Morell, *Beyond Superfailure: America's Toxic Policy for the 1990s* (Boulder, Colo.: Westview, 1992), chap. 5. See also Joel S. Hirschhorn and Kirsten U. Oldenburg, *Prosperity without Pollution: The Prevention Strategy for Industry and Consumers* (Von Nostrand Reinhold, 1991).

14. William T. Gormley, Jr., *The Politics of Public Utility Regulation* (University of Pittsburgh Press, 1983).

15. Barry G. Rabe, "Legislative Incapacity: Congressional Policymaking and the Case of Superfund," *Journal of Health Politics, Policy and Law*, vol. 15 (Fall 1990), pp. 571–90.

16. Jonathan M. Moses, "Insurer Payouts over Superfund Flow to Lawyers," *Wall Street Journal*, April 24, 1992, p. B1.

17. John H. Cushman, Jr., "Administration Plans Revision to Ease Toxic Cleanup Criteria," *New York Times*, January 31, 1994, p. A1; and Timothy Noah, "Clinton Today Begins an Attempt to Clean Up Big Toxic-Waste Problem: Superfund Law Itself," *Wall Street Journal*, February 3, 1994, p. A16.

18. Barry G. Rabe, *Fragmentation and Integration in State Environmental Management* (Washington: Conservation Foundation, 1986). See also "Integrated Pollution Control: A Symposium," *Environmental Law*, vol. 22, no. 1 (1992), pp. 1–348.

19. V. O. Key, *Public Opinion and American Democracy* (Knopf, 1961), p. 555. See

also Benjamin I. Page and Robert Y. Shapiro, *The Rational Public: Fifty Years of Trends in Americans' Policy Preferences* (University of Chicago Press, 1992). Page and Shapiro provide considerable insight into the promise of collective deliberation in public policy, particularly in chapter 10.

20. Mary Ann Glendon, *Rights Talk: The Impoverishment of Political Discourse* (Free Press, 1991), p. 143.

21. Chris Zeiss, "Directions for Engineering Contributions to Successfully Siting Hazardous Waste Facilities," paper prepared for the September 1993 conference on Siting Hazardous Waste Facilities: Politics and Approaches for the 1990s, University of British Columbia, pp. 35–36.

Index

Alabama: Emelle landfill, 43; waste disposal and treatment capability, 20; waste import, 20, 49

Alberta: abandoned dump cleanup, 164; Alberta Environment, 66–67, 68, 72–73, 74, 76; Alberta Heritage Savings Trust Fund, 62; Alberta Rural Education and Development Association, 67; Alberta Special Waste Management Corporation, 72, 76; Alberta Special Waste Service Association, 85; Alberta Waste Materials Exchange, 85; burden sharing, 79–80; community meetings, 66–67; crown corporation, 61, 72, 74, 75, 78; economy, 62; Environmental Council, 66; hazardous waste disposal before 1987, 63; Hazardous Waste Management Committee, 64, 65–66; hazardous waste quantity, 62; Hazardous Waste Task Force, 66, 67; market-based approach to siting in, 63–64; Ministry of Environment, 61; protest groups, 63–64, 68–69; public participation in waste facility siting, 64–71; recycling, 84–86; regional waste storage facilities, 83–84, 87; voluntary approach to siting in, 58–89; as voluntary siting model, 87–88, 90, 91–100, 106, 110–13, 121, 126, 133; waste export, 26, 63, 81; waste reduction, 84–86. See also Swan Hills Special Waste Treatment Centre

Alternative dispute resolution, 9–10, 36

Arizona, siting in, 56

Atomic Energy Act of 1946 (Canada), 130

Atomic Energy Control Board (AECB), 130, 131–32, 133, 135, 139

Atomic Energy of Canada Limited (AECL), 130, 131

Augusta Environmental Strategies Committee (Michigan), 44

Barbee, Tom, 107, 108, 109, 110

Beaudette, Florent, 99

Brandeis, Louis, 150

British Columbia: environmental regulation, 39, 40; Hazardous Waste Advisory Committee, 39; Hazardous Waste Management Coalition, 41; siting attempts in, 23, 29, 39–42, 151; waste export, 26, 39, 42

British Columbia Waste Management Act, 39

Browning-Ferris Industries, 16

Bruce-Viking Agricultural Protection Association (Alberta), 67

Bullard, Robert D., 21

Burden sharing, 56, 59–60, 157–61; in Alberta, 79–80, 83–84; in California, 91, 118–21; for LLRW disposal, 138, 139; in Manitoba, 101–06; regional agreements, 160–61; restrictions on in United States, 80, 103, 159, 160

Bureau of Engraving, 118

California: Appeals Board, 120; burden sharing, 91, 118–21; Department of Health Services, 118, 120; LLRW management, 142;
waste export, 121; waste management,
118–21; waste reduction, 120–21
CAMECO, 132
Campbell, Carroll, 20
Canada: decentralized regulatory policy,
150–52; environmental policy, 23–24;
guidelines for on-site disposal, 17; hazard-
ous waste policy similarities to U.S., 1,
22–27; Nimby syndrome in, 23; political
culture, 23–24
Candu 3 nuclear reactor, 129
Carey, Hugh L., 51
Carlson, Arne, 117
Central Compact, 32, 143, 144, 145
Central Interstate Compact, 32
Chalk River Nuclear Laboratories, 131, 136
Champion, Jacquie, 74–75
Charlottetown Constitutional Accord, 65
Chemical Waste Management, 43
Ciba-Geigy, 162
Citizens, 138; capable of collective delibera-
tion, 7–8; environmental decisionmaking
participation, 64, 65; political interest,
14–15; public distrust, 24, 71, 135. See also
Protest groups; Public participation
Clean Harbors, 37
Clinton administration, 159
Coalition for Safe Waste Management (Mas-
sachusetts), 38
Communities, 168; disposal technology selection, 138; exploitation concerns, 59, 91,
109, 139, 146; Massachusetts advisory committees, 36; rejection of disposal sites, 33;
siting meetings in Alberta, 66–67; volunteering for waste disposal sites, 133; waste
disposal siting concerns, 135, 136
Community liaison groups: in Alberta, 66–67;
in Manitoba, 92–93, 96; in Port Hope, Ontario, 134, 135
Compacts, for LLRW disposal, 138, 140–43,
153, 160; regulatory approach of, 142–43
Compensation packages, 36, 37, 76, 139, 145;
in Minnesota, 111, 112; in Quebec, 123;
for Swan Hills, Alberta, 76–79
Concerned Citizens for Ecology and Environment (Manitoba), 95–96
Concerned Citizens for Nebraska, 144
Conklin Company, 118
Connecticut: community protest, 142–43;
Hazardous Waste Management Service,
142–43; voluntary approach to siting, 147;
waste export, 26
Cookson, Jack, 67

Cronin, Thomas E., 155
Crown corporations, 72, 132, 135, 163; Alberta, 61, 72, 74, 75, 78; Manitoba, 92, 95,
96–99; Ontario, 46, 47

Davis, Gary, 37
Deep River, Ontario, 136
Deep-well injection, 17, 161
Disposal. See Hazardous waste disposal
Duerksen, Christopher, 45
Duguid, Terry, 99
Dukakis, Michael S., 35

ECOFLO, 107–10
Eldorado Resources Limited, 132, 139
English, Mary, 37
Envirochem Group, 40, 41
Environment, Ministry of the, 24, 41
Environmental Facilities Corporation, 51
Environmental organizations, 69–70, 79
Environmental Protection Act of 1988 (Can-
ada), 12, 25
Environmental Protection Agency (EPA), 11,
25, 71, 102, 108, 151, 166; moratorium on
incineration construction, 12; RCRA en-
forcement, 13
Environment Canada, 25, 81, 151–52
Envotech Limited Partnership, 43, 44
Esman, Milton J., 24
Europe: environmental policy in, 22; waste
management facility siting, 149; waste man-
agement technologies, 162
Exxon, 162

Federalism, 150, 152; Canadian and American compared, 25–26, 152–53
Filmon, Gary, 97
Florida: Department of Environmental Regulation, 49, 50, 51; regulatory approach to
siting in, 24, 49–51
Friends of the Environment in Swan Hills, 69

Georgia: regulatory approach to siting in,
55–56; waste import, 20
Gimello, Richard J., 55
Glendon, Mary Ann, 9, 168
Gormley, William T., Jr., 45, 71
Gravure Graphics, 105–06
Greenberg, Michael R., 21
Greenpeace, 69
Gruber, William, 106
Guilford County Hazardous Waste Task
Force, 108

Haldiman-Norfolk Organization for a Pure
Environment (Ontario), 47

Hazardous waste: benefits from, 1–2; defined, 10–11; education programs, 86; as interstate commerce item, 80, 103; oil, 85; PCBs, 63, 81, 82, 102; pharmaceutical goods, 85; quantities, 1, 11–12, 15, 62; regional storage facilities, 83–84, 87; spills, 18–19, 82. *See also* Waste management; Waste reduction; Waste transport
Hazardous waste disposal siting. *See* Siting
Hazardous waste imports and exports. *See* Waste transport
Hazardous Waste Treatment Council, 26
Heiman, Michael, 104
Help End Landfill Pollution (Alberta), 164

Ilgen, Thomas L., 27
Incineration, 12, 161
Indiana, waste import, 20
Inform, 161–62
Institutions: change in, 7, 71–75, 96–99, 113; public confidence in, 8, 14–15, 71, 163
International Environmental Policy Coalition, 26
International Joint Commission, 137
Interstate compacts. *See* Compacts
IT Corporation, 16, 37, 39

Johnson, Tom, 113

Keep Freetown Hazard Free (Massachusetts), 37
Key, V. O., 167
Kinetic Contaminants Canada, 63–64
Kowalski, Ken, 68
Krawetz, Natalia M., 81

Laidlaw, 16
LaVo, Carl, 55
League of Women Voters (Nebraska), 144
Lipset, Seymour Martin, 23
LLRW. *See* Low-level radioactive waste
Louisiana, waste import, 20
Love Canal, 29
Low-level radioactive waste (LLRW), 128; Canadian management, 131–39; federal regulation of, 129–31; siting process in Nebraska, 143–47; siting deadlock, 128–29; sources, 130, 137, 139, 143, 144; U.S. management, 140–43
Low-Level Radioactive Waste Policy Act of *1980*, 130, 140
Lynn, Frances M., 43, 107, 108

McQuaid-Cook, Jennifer, 74–75
Magorian, Christopher, 45
Manitoba: community meetings, 92–93; crown corporation, 92, 95, 96–99; hazardous waste quantity, 92; Manitoba Environmental Centre, 92, 94; Manitoba Hazardous Waste Management Corporation, 96–99; public participation in waste facility siting, 94–96, 98; recycling, 104–06; regional storage facilities, 103–04; voluntary approach to siting in, 48, 91–101; waste export to United States, 26; waste import control, 101–03, 114; waste reduction, 104–06
March, James G., 8
Market-based approach to siting, 7, 30–34, 45; in Alberta, 63–64; in British Columbia, 39–42, 151; compensation, 56, 76; failures, 46, 49, 76; in Massachusetts, 34–38; in Michigan, 43–44; in North Carolina, 42–43. *See also* Siting
Massachusetts: Coalition for Safe Waste Management, 38; Department of Environmental Protection, 37; environmental regulation, 38; Hazardous Waste Facilities Site Safety Council, 35; Hazardous Waste Facility Siting Act of *1980*, 35; market-based approach to siting in, 34–38; waste export, 19, 26, 35, 126
Matheny, Albert R., 45
Mazmanian, Daniel A., 17, 71
Mennie, Don, 100
Michigan: LLRW siting process in, 143; market-based approach to siting, 43–44; waste export, 26; waste import, 44
Mick, Lorne, 75
Midwest Compact, 143
Milan Citizens Against Toxic Substances (Michigan), 44
Minnesota: compensation packages, 111, 112; disposal timetables, 11; Minnesota Technical Assistance Program (MnTAP), 116, 117; Office of Waste Management, 113, 115–16, 117; pollution prevention, 116, 117; protest groups in, 111–12; regulatory approach to siting in, 111; Toxic Pollution Prevention Act, 116; voluntary approach to siting in, 91, 110–18; waste export, 26; waste import control, 114; Waste Management Act, 111, 117; Waste Management Board, 111, 112, 113, 114, 115; waste reduction, 116–18
Morell, David L., 17, 45, 119–20
Morone, James A., 155

National coal policy project, 10
National Governors Association, 140, 160
Nebraska: Department of Environmental Control, 144; Nebraska Citizens Advisory

Committee (NCAC), 144, 145; voluntary approach to siting in, 143–47
Newalta Corporation, 84
Newcastle, Ontario, 137, 138
New Jersey: Department of Environmental Protection, 54; Hazardous Waste Advisory Council, 54–55; hazardous waste explosions, 15; Major Hazardous Waste Facilities Siting Act, 54; protest groups, 55; regulatory approach to siting in, 24, 54–55; waste disposal costs, 89; waste export, 19, 26, 54, 126, 142; waste import, 20
New York: Department of Environmental Conservation, 35, 51; Love Canal, 29; protest groups, 52; regulatory approach to siting in, 24, 29, 51–52; waste export, 126; waste import, 35
Nienaber, Jeanne, 71
Nimby, 2–3, 167; in Canada, 23; protest outcomes, 56–57, 58; as public health problem, 14; public involvement and, 14–15. See also Protest groups
Norrie, Bill, 94
North Carolina: ECOFLO facility, 107–08, 109; Hazardous Waste Facility Siting Commission, 52–53; market-based approach to siting in, 42–43; public participation in facility siting, 107–09; regulatory approach to siting in, 52–54; Solid and Hazardous Waste Management Branch, 42; voluntary approach to siting in, 106–10; waste export, 20, 106
Not-In-My-Backyard syndrome. See Nimby
Nuclear power plants, 130, 131

Office of Technology Assessment, 161
Ohio, waste disposal and treatment capability, 20
Olsen, Johan P., 8
Ontario: crown corporation, 47; low-level radioactive waste in, 131, 137, 139; Ontario Hydro, 130–31; Ontario Waste Management Corporation (OWMC), 24, 46–48, 50; public distrust in, 24, 135; regulatory approach to siting in, 19, 22, 24, 46–49, 151; Siting Task Force, 133–35, 137, 139; Toxic Waste Research Coalition, 48; voluntary approach to siting for LLRW disposal, 48–49, 132–39, 154; waste export, 26
Oregon, waste import, 81
Ostrom, Elinor, 7

Participatory democracy, 64–65, 154–55
Pavelich, Joseph, 113
Pennsylvania, waste import, 20
People for Progress (Nebraska), 146
Peretz, Jean, 45

Policy professionals, 73–74, 98, 113
Pollution Probe, 162
Port Hope, Ontario, 137; harbor, 137, 139; radioactive wastes in, 131–32, 136; voluntary site for LLRW disposal, 136–37, 138, 139
Protest groups: in Alberta, 63–64, 68–69; in Arizona, 56; in British Columbia, 41; in Connecticut, 142–43; in Florida, 50–51; in Massachusetts, 36–38; in Michigan, 44, 143; in Minnesota, 111–12; in Nebraska, 146; in New Jersey, 55; in New York, 52; in Ontario, 47–48. See also Citizens; Nimby; and names of specific groups
Public distrust, 14, 24, 71, 127, 135, 163
Public hearings. See Public participation
Public information, 59, 66, 95, 108, 112, 138
Public participation, 59, 64–71, 94–96, 107–09, 123, 154–55, 163. See also Citizens; Participatory democracy; Protest groups

Quebec, 26; compensation packages, 123; hazardous waste management, 124–27, 151; hazardous waste quantity, 125; illegal dumping in, 125, 126; post-siting breaches of promise, 91, 124–26, 127; public distrust, 127; public participation in siting, 122–23; regulation enforcement lacking, 125; regulatory approach to siting, 122; waste export, 19, 82, 126; waste import, 19, 126
Quirk, Paul J., 6

Radioactive waste contamination, 131–32
RCRA. See Resource Conservation and Recovery Act of 1976
Recycling, 11, 15, 84–86, 104–06
Regulatory approach to siting, 29, 31–33, 56; in Arizona, 56; by interstate compacts, 142–43; in Florida, 49–51; limitations of, 44–46; in New Jersey, 54–55; in New York, 51–52; in North Carolina, 52–54; in Ontario, 46–49; voluntary process compared, 4–6. See also Siting
Regulatory policy, 24–25, 45, 150–52, 165–66
Resource Conservation and Recovery Act of 1976 (RCRA), 27, 159; enforcement, 13; hazardous waste definition, 10–11; as litigation stimulus, 12
Richards, Alun, 98
Ristoratore, Mario, 25, 26
Ryz, Marvin, 102

Sargent, Francis, 38
Saskatchewan: siting approach, 23; waste export, 26
Save Boyd County (Nebraska), 146
Simpson, Kenneth J., 75

Siting, 148–49; breaches of promise after, 91, 124–26, 127, 139, 146; centralized versus decentralized processes, 29–32; collective deliberation process, 7–8; community concerns, 135; compensation for, 141–42; constraint mapping, 66; cooperative agreements on, 3; cooperative versus coercive agreements, 4–6; economic benefits of disposal sites, 76–78; by eminent domain, 44; expert judgment and, 45; institutional reform needed, 71–75, 96–99, 163; Native American lands, 21; before 1970s, 28–29; political leadership qualities for, 44; by preemption, 44–56; societal costs of gridlock, 13–14; socioeconomic inequities, 21–22; traditional strategies, 3, 6–7; unified versus divided government and, 24. See also Market-based approach to siting; Regulatory approach to siting; Voluntary approach to siting

South Carolina, 26; waste disposal and treatment capability, 19–20; waste import, 20, 49, 142

Southern California Waste Management Forum, 14

Stablex Canada, 40, 121–26

States, and interstate LLRW management compacts, 138, 140–43, 153

Stop-IT (Massachusetts), 37

Superfund program, 12, 13, 71, 102, 151, 152–53, 159; capacity assurance plans, 13, 25, 49, 159; reform needed, 164–65

Supreme Court. See U.S. Supreme Court

Supreme Court of Canada, 24

Swan Hills Special Waste Treatment Centre (Alberta), 61–62, 70, 86, 164; community benefits from, 76–79; community meetings, 70, 74–75; costs, 88–89; import control, 80–83; public access to, 70–71; as voluntary siting model, 87–88, 90, 91–100, 106, 110–13, 121, 126. See also Alberta

Syncrude Canada, 85

Tanner act (California), 119–21

Technology, and waste management industry, 16–17, 161–62

Tennessee, waste import, 20

Thompson, Fred, 27

Thornton, Susan, 113

Toxic waste. See Hazardous waste

Transportation of Dangerous Goods Act (Canada), 12

Transportation of waste. See Waste Transport

Underground injection, 17, 161

United Citizens Against Pollution (Florida), 50–51

U.S. Supreme Court, 11, 80, 103, 140, 159, 160

US Ecology (Nebraska), 144–47

Virginia, waste import, 20

Voluntary approach to siting, 3, 59–60, 90–91, 167–69; in Alberta, 58–89; Alberta as model for, 87–88, 90, 91–100, 106, 110–13, 121, 126, 133; breach of promise protection, 139; coercive process compared, 4–6; compensation packages, 139; in Connecticut, 147; exploitation concerns, 59; in Manitoba, 48, 91–101; in Minnesota, 91, 110–18; in Nebraska, 143–47; need for, 149–50, 155–56; in North Carolina, 106–10; in Ontario, 48–49, 132–39; political administration changes and, 138; public education, 59, 138; public participation, 59, 64–71, 94–96, 107–09, 163. See also Siting

Washington, LLRW imports, 142

Waste Management. See WMX Technologies

Waste management: Canadian and U.S. similarities, 1, 22–27, 149; incineration, 12–13, 161; on-site, 17–18; recycling, 11, 15, 84–86, 104–06; safety improvements, 15; technologies, 15–16, 161, 162; transborder waste trading, 19–20, 26–27; underground injection, 17, 161; waste management industry and, 16–17; waste reduction, 84–86, 104–06, 139, 160, 161–62

Waste management industry, 16, 26–27; disposal technology and, 16–17; Native American site negotiations by, 21–22

Waste reduction, 11, 139, 160, 161–62; in Alberta, 84–86; in Manitoba, 104–06; in Minnesota, 115–18

Waste transport, 18–20, 49, 80–83, 102, 121, 142, 160; transborder waste trading, 19–20, 26–27; vehicles for, 84

Weston, Roy F., Inc., 50, 51

Williams, Bruce A., 45

WMX Technologies, 16, 43